Major soils and soil regions in the Netherlands

Major soils and soil regions in the Netherlands

H. de Bakker

Soil Survey Institute, Wageningen, the Netherlands

Springer-Science+Business Media, B.V

ISBN 978-94-009-9986-2   ISBN 978-94-009-9984-8 (eBook)

DOI: 10.1007/ 978-94-009-9984-8

Dr W. Junk

© Springer Science+Business Media Dordrecht 1978

Springer-Science+Business Media, B.VDr W. Junk B.V. Publishers,

The Hague, the Netherlands, Boston, U.S.A., London, U .K. and

Centre for Agricultural Publishing and Documentation, Wagellingen, the Netherlands, 1978.

Design: Pudoc, Wageningen

Softcover reprint of the hardcover 1st edition 1978

'Man and soil were made to be partners on the earth. . . man would take his partner for better not worse, for richer not poorer, not for sickness but health until death do them join.'

H. S. Gibbs, 1960

Contents

Foreword

Soil science in the Netherlands puts strong emphasis on the relationship between parent material and soil formation, and between physiographic conditions and land use. This approach, developed by the late Professor Dr C.H. Edelman (1903-1964), is quite understandable in a country where soils have to a great extent developed from alluvial and aeolian materials of recent geological origin. Dutch soil scientists have paid much attention to pedogenesis in fresh sediments, known as 'initial soil formation' or 'ripening', and to groundwater as a soil forming factor. Furthermore, human influence on soil genesis, in this land of man-made soils, has been thoroughly investigated.

'Major soils and soil regions in the Netherlands' clearly reflects these specific features of soils work in this country. In his book, Mr H. de Bakker, Head of the Soil Classification Section of the Netherlands Soil Survey Institute (Stichting voor Bodemkartering) addresses the special interests of foreign soil scientists and of students of earth sciences. The author examines representative soil profiles not only with respect to the Netherlands' system of soil classification – of which he is the co-author – but also in terms of some of the major classification systems used in other countries. It appears that a well characterized soil profile may find a very different place in various classifications and that even for a trained specialist it is often difficult to arrive at a specific determination within a given system. This difficulty applies especially to soils from recent deposits which constitute a major proportion of the soil pattern of the Netherlands and have been intensively studied in this country. Elsewhere they are often lumped together in broader units with no provision being made for detailed subdivisions.

Although Mr de Bakker's interpretation may not always be agreed upon by other colleagues, the excellent descriptions and this remarkably illustrated review of 'Netherlands' soils in colour' should promote further correlation. It is hoped that this work will lead to a better understanding and a streamlining of 'pedological language' used to compare soils in different parts of the world and to assess their suitability for various types of land use.

R. Dudal, Director
Land and Water Development Division
Food & Agriculture Organization of the United Nations
Rome, Italy

Introduction

The thirty two colour plates of Dutch soils around which this book is built, were originally published in a series of articles in a Dutch periodical (Tijdschrift der Koninklijke Nederlandsche Heidemaatschappij) between 1964 and 1966. Each article contained a colour plate with an accompanying profile description, an account of the geography of the soil and an appraisal of the analytical data together with an oblique aerial photograph in black and white of a suitable landscape with explanation.

The main purpose of these articles has been to introduce the system of soil classification for the Netherlands and its new nomenclature (De Bakker & Schelling, 1966), which is currently being used on the 1 : 50 000 Soil Map of the Netherlands (Stichting voor Bodemkartering, 1964- ..).

Recently these articles were revised and published in a book (De Bakker & Edelman-Vlam, 1976). Like the articles it was written for students, teachers (the plates are made available as slides) and scientists with a knowledge of the geography and history of the Netherlands.

In printing the plates for the Dutch book, it was thought worthwhile to use the same colour plates for an English edition. It was not thought practical to translate the original, so a new text has evolved. This text not only discusses the soils but also the environmental conditions and seven soil regions in the Netherlands; these chapters are designed to give the reader a background knowledge of the Netherlands, its history of reclamation, settlement and agriculture. The main part of the book, the chapter 'Soils', gives a detailed appraisal of the soils, their genesis, classification, location and development. Since the American Yearbook of Agriculture (USDA, 1938), the Soil Map of Europe (Dudal et al., 1966) and the *Bestimmungsbuch* of Kubiëna (1953) soil classification has developed considerably: e.g. Soil Taxonomy (SSS, 1975), the Soil Map of the World (FAO, 1974) and Mückenhausen et al., 1977). In view of this development ample attention is devoted in the chapter 'Soils' to the place of the discussed soils in several systems of soil classification (cf. p. 60).

Environmental conditions

The Netherlands (Fig. 1) is one of the small countries in northwestern Europe. It is situated between the North Sea in the north and Belgium in the south (between latitudes 51° N and 54° N), and between the North Sea in the west and the Federal

Fig. 1. Location of the discussed soils.

Republic of Germany in the east (between longitudes 3° E and 6° E). The total area of the Netherlands (1976) is 34 000 km² (exclusive of water). About 11% is built-up area (residential districts, industrial areas, roads, airports, parks); 9% is under forest, of which about half is coniferous forest; about 75% is agricultural land (of which 61% is grassland, 33% is arable land and 5.5% is horticultural land); the remaining area (5%) consists of heathlands, coastal and inland dunes, coastal marshes, reed marshes, these formerly were called waste land, but today they are very valuable for outdoor recreation and for nature conservancy.

By enclosing new polders the area has been increased by about 5000 ha per year in the last two decades, but urban expansion has claimed yearly 8000 to 10 000 ha in the same period.

There are 13.8 million inhabitants (31 December 1976). Of the economically active population 7% work in agriculture and fisheries, 43% in industry, 26% in commerce, banking and transport, and 24% in other services.

Some detailed information on environmental conditions is given in terms of the classical factors of soil formation (Jenny, 1941, 1961) in the following five sections entitled: Parent material, Climate, Time, Topography, and Biotic factors.

Parent material

Nearly all mineral soils in the Netherlands are developed from clastic sediments, with textures ranging from fine sands to clays. They may be aeolian (loess, cover sand, coastal and inland dune sand), fluviatile (sediments of the Rhine and the Meuse), marine (tidal sediments of the North Sea and its inlets), or glacial (glacial till and fluvioglacial). The only soils derived from solid rock in the Netherlands, are the rendzina soils developed from Cretaceous chalk outcropping on slopes in the south of Limburg Province (soil L3, p.189). The parent material of the organic soils ranges from eutrophic wood peat to oligotrophic Sphagnum moss peat.

Figure 2 shows a generalized distribution of the various kinds of parent material in the Netherlands, differentiated according to geological age, texture and origin. In Table 1 particle-size distributions are given for two samples from each of the ten mapping units delineated. The first example is typical and the second a common variant.

Fine sands of coastal and inland dunes. The first mapping unit comprises coastal dunes and inland dunes, represented in Table 1 by Samples 1a and 1b respectively. Both sands are well sorted, the former a little more thoroughly than the latter. Inland dune sand, being derived from the Pleistocene cover sand, is always non-calcareous. The coastal dune sand north of the gap in the dunes on the mainland (south of Den Helder) and on the Frisian Islands is also non-calcareous. The sands of the south are calcareous nearly everywhere, for several reasons, one of them being the difference in origin (Eisma, 1968). Soil A3 (p.145) exemplifies a soil on an inland dune.

Loamy materials of marine origin. Both Samples 2a and 2b are from the marine areas indicated as the second mapping unit of Figure 2. Sample 2a is typical of the recent, calcareous young polders (Fig.5), the latter is less extensive and can be found in the transition zone between the marine area and the adjacent peat lands

(Fig.8). When comparing Sample 2a with 4a (a sample from a natural levee from the Rhine) it is obvious that the marine sediment is more silty and has finer sand than the fluviatile sediment, these being typical differences between the two sedimental environments. On the other hand fine-textured river sediments (4b) have a smaller particle-size range than similar marine sediments (2b). Examples of marine soils developed in sediments such as characterized by Sample 2a are soil M6 (p.85) and LP3 (p.113); and of Sample 2b: soil M3 (p.73) and the topsoil of soil M7 (p.89).

Sands of marine origin. The marine sands are illustrated by the Samples 3a and 3b. The first is a normal sea sand, often found in the subsoils of marine polders (soil M4, p.77), or rarely as whole soils, such as soil M5 (p.81). The sand in the western part of the Lake-Yssel polders is much finer (Sample 3b); in the east the polders are part of the embanked foreset beds of the delta of the river Yssel. The sand here is much coarser (40-50% between 150 and 210 μm. Examples of soils developed from these coarse Lake-Yssel sands are not presented in this book.

Loamy and clayey materials of fluviatile origin. The fluviatile sediments are exemplified in Table 1 by Samples 4a (from a natural levee) and 4b (from a backswamp), and by soil F3 (p.101) and F1 (p.93) respectively. See also the previous comparison between marine and fluviatile sediments.

Organic materials. No particle-size analyses are given for organic materials, as these are meaningless because of the high organic-matter content.

In Western Europe peats are conventionally subdivided into high moors and low moors (Mückenhausen et al., 1977, p. 167-169; Kubiëna, 1953, p. 109 and 134). This subdivision is based on the height of the peat relative to the surrounding land.

During its upward growth a peat changes from a level low moor (fen), dependent on both ground water and rain water, into a domed high moor (raised bog) solely dependent on precipitation. The acidity and the fertility of the sites are determined by these differences in moisture regime, which consequently also determine the peat-forming vegetation.

In my opinion we should discard the terms 'high' and 'low' because in many places in the Netherlands high moors are found below sea level (former raised bogs drowned by the post-glacial rise in sea level and covered with marine sediments; Figure 8). A more precise description might be oligotrophic and eutrophic peat, or ombrogenic and topogenic peat or perhaps more simply distinction could be made by plant remains, Sphagnum moss peat, wood peat, etc. Soil LP1 is an organic soil developed from wood peat; the soils discussed in p.114-133 are from conventional high moor districts and all have oligotrophic peat (mostly Sphagnum moss peat) in their sola.

Loamy materials derived from loess. The soil parent materials in Belgium and Germany have not been shown in Figure 2, except for the loess, which has been put in to show that the Dutch loess region is part of the West European loess belt. Its location at the northern extremity of this belt explains the rather low clay content of Sample 6a, a calcareous C horizon. Sample 6b is from a clay-enriched Bt horizon. If the sediment is thick enough, the depth of decalcification is 2.5-3 m, the original lime content being 10-20%. Two soils developed from loess are discussed in this book: soil L1 (p.181) and soil L2 (p.185).

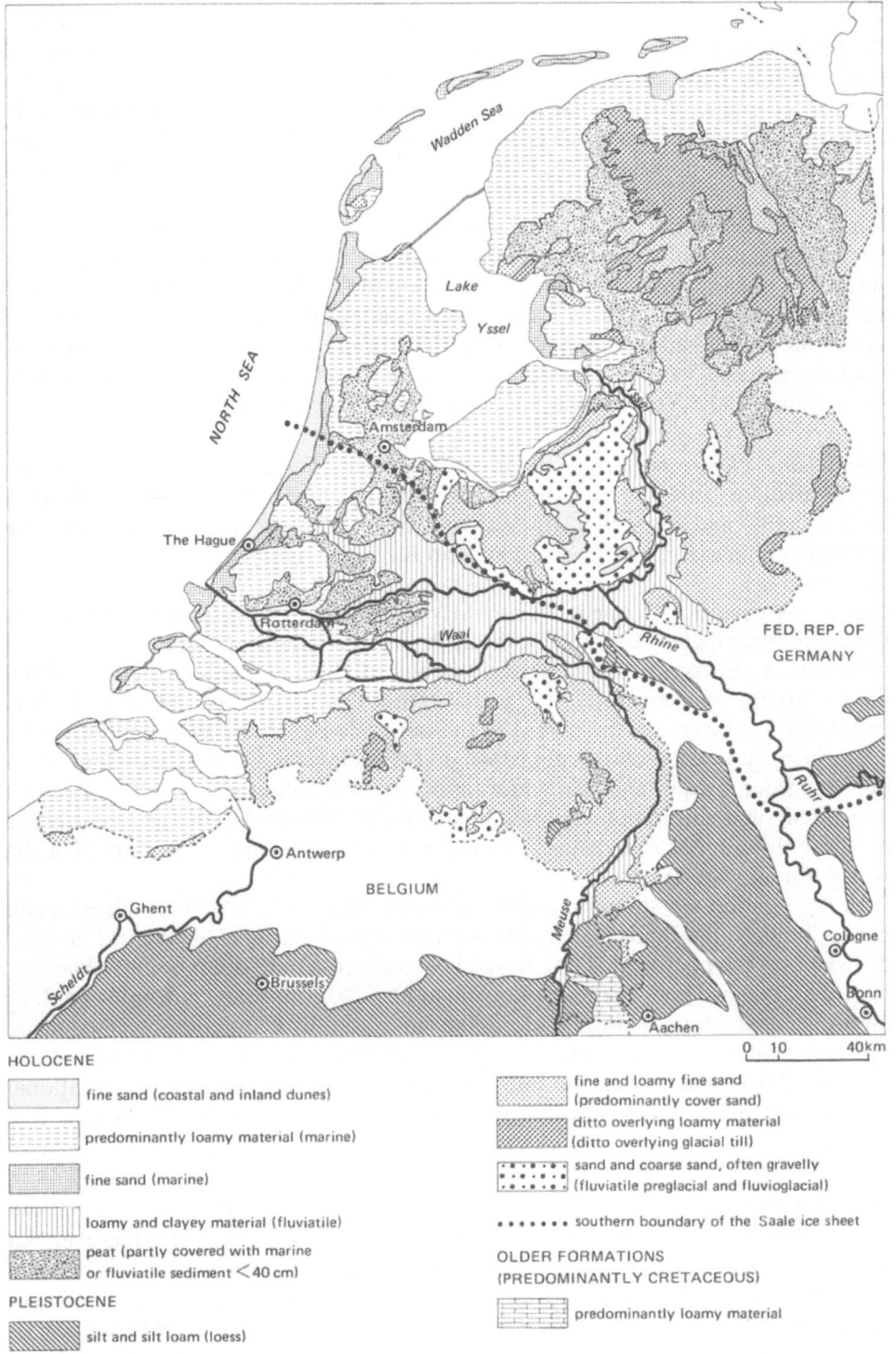

HOLOCENE

	fine sand (coastal and inland dunes)
	predominantly loamy material (marine)
	fine sand (marine)
	loamy and clayey material (fluviatile)
	peat (partly covered with marine or fluviatile sediment < 40 cm)

PLEISTOCENE

	silt and silt loam (loess)

	fine and loamy fine sand (predominantly cover sand)
	ditto overlying loamy material (ditto overlying glacial till)
	sand and coarse sand, often gravelly (fluviatile preglacial and fluvioglacial)
••••••	southern boundary of the Saale ice sheet

OLDER FORMATIONS (PREDOMINANTLY CRETACEOUS)

	predominantly loamy material

Fig. 2. Parent material and surface geology in the Netherlands. In Belgium and in the Federal Republic of Germany only the loess is indicated.

Table 1. Particle-size analyses of parent materials in the Netherlands

Parent material groups	No.	Particle-size distribution (%) of parent material (μm)																	
		<2	2-4	4-8	8-16	16-25	25-35	35-50	50-75	75-105	105-150	150-210	210-300	300-420	420-600	600-850	850-1200	1200-1700	gravel
fine sands of coastal and in-	1a	1.5	0.3	0.2	0.1	0.1	0.1	0.0	0.2	0.7	12	44	32	8	0.7	–	–	–	–
land dunes	1b	0.0	0.0	1.0	0.3	0.0	0.0	1.0	4	12	27	37	16	2.5	0.3	–	–	–	–
loamy materials of marine	2a	19	1	3	3	4.5	6	14	15	20	14	0.5	0.2	0.1	–	–	–	–	–
origin	2b	38	8	11	9	5	4	3	6	9	7	0.4	0.1	0.1	–	–	–	–	–
sands of marine	3a	1	0.4	0.2	0.5	0.5	0.4	2.5	12	23	36	18	5	0.5	0.2	0.1	–	–	–
origin	3b	6.5	1	0.5	0.5	1	2	2	20	44	21	0.3	0.2	0.1	–	–	–	–	–
loamy and clayey materials	4a	18	2.5	3	7	3.5	3.5	5	3.5	6	9	14	16	5	2.5	1	0.3	–	–
of fluviatile origin	4b	48	8	10	11	6	9	7	1	0.4	0.1	0.1	0.1	0.1	0.1	0.1	–	–	–
loamy materials derived	6a	15	0.5	3	6	15	54	3.5	3	0.3	0.1	0.1	0.0	0.1	–	–	–	–	–
from loess	6b	21	4.5	5	12	13	29	3	0.4	0.2	0.1	0.3	–	–	–	–	–	–	–
fine and loamy fine sands	7a	4	0.0	0.0	0.0	0.5	0.2	7	10	17	34	18	7	1.5	0.5	0.2	0.2	0.1	–
derived from cover sand	7b	4	0.2	0.2	1	2	4.5	20	11	13	21	15	6	1.5	0.5	0.3	0.1	–	–
glacial tills	8a	18	2.5	2	3	2	2.5	3	4	11	14	14	12	7	2.5	1.5	1	0.3	7
	8b	28	2	3	2.5	2	2.5	4	4.5	9	8	13	9	4.5	3.5	2	1.5	1	3
sandy and gravelly materials of	9a	1.5	0.1	1	1	1	0.5	6	2	2	8	15	18	21	12	7	3	1	14
ice-pushed ridges	9b	3	0.4	0.0	0.2	0.6	1	2	1	2	10	19	22	15	12	7	3	2	22
loamy materials derived from	10a	42		13			33		4.5	2	2				5				–
Cretaceous rocks	10b	25		8			18		21	11	6				12				–

Sandy materials derived from cover sands. Cover sand (the seventh mapping unit on Figure 2) is a widespread aeolian deposit mainly of late Weichsel age. In places it is several metres thick. Elsewhere it veneers older sandy sediments, for example preglacial coarse river sediments and glacial outwash. It is non-calcareous, at least in its upper part; it is believed to have been calcareous when originally deposited on barren floodplains extending across the dry North Sea basin (see also p.174). Gley soils such as soil P6 (p.177) are often calcareous at 1-2 m depth, but podzolized soils are non-calcareous in the deep C horizons. Two Samples are entered in Table 1; 7a is an example of a 'normal' (i.e. often encountered) cover sand with little silt fraction, 7b is less extensive and its silt content illustrates its relation to loess. Soils P1 (p.157), P3 (p.165), P6 (p.177) and the subsoils from the district of the cut-over raised bogs (p.114-133) are typical of the soils formed on cover sands.

Thin cover sands overlying glacial till. In the north of the country, cover sand overlies glacial till (the eight mapping unit of Figure 2, Sample 8a in Table 1). Although its clay content is rather low (mostly between 15 and 25%) it is a compact and cohesive material. Therefore it is excellent for dike-building. Glacial till outcrops on the sea bed near the enclosure dam between the Wadden Sea and Lake Yssel and it was dredged and used to build the dam in 1932. The glacial till of the two small areas in the east is derived from Tertiary clay: this material is more clayey than the till further north (Sample 8b in Table 1). In the south the cover sand overlies loess and an early-Pleistocene fine-textured fluviatile sediment.

Sandy and gravelly material of ice-pushed ridges. Ice-pushed ridges and fluviogla-cial plains, Samples 9a and 9b, form the ninth mapping unit. The ridges to the north of the dotted line on Figure 2 are formed of middle and lower Pleistocene coarse-textured fluviatile sediments, shaped into low hills by the Saale ice sheet at the end of the middle Pleistocene. The fluvioglacial plains date from the same period. The small areas of this mapping unit to the south have a similar origin but were not affected by the ice sheet. The sands in Samples 9a and 9b are ill-sorted compared with the well-sorted sands of aeolian parent materials such as the dune sands and the cover sands (Samples 1a, 1b, 7a and 7b), furthermore they have a long 'tail' at the coarse side.

Loamy materials derived from Cretaceous rocks. The older sediments are upper-Cretaceous and are not covered with loess (Fig.32, 33 and 34). In Table 1 this mapping unit is represented by a sample of the weathering product of a chalk (Sample 10a), formerly called 'sticking earth' (Breteler, 1958), and by a glauconite clay (Sample 10b). No detailed particle-size analyses of these parent materials are available. This mapping unit includes a part of the Late-Tertiary plain on the boundary between the Netherlands, Belgium and Germany (Van den Broek & Van der Waals, 1967) with soils developed in clay-with-flints (Hodgson, 1967, p.7 and 8; Hodgson et al., 1967) or *argile à silex* (Duchaufour, 1965, p.162). No sample of this material has been included Table 1.

Climate

The climate of the Netherlands is a Cfb-climate according to Köppen's classification. The winters are mild, even the temperature in the coldest month is above 0 °C; in summer there are four months with a mean temperature over 10 °C, and the precipitation is evenly distributed over the year (Table 2).

Due to differences in evaporation there is a precipitation deficit during the growing season. For this reason the water supply of crops depends for a major part on the availability of soil moisture. Shallow soils (such as soils M4 and F2) or shallowly rootable sandy soils (such as soils M5, P1 and P3) suffer from drought in dry summers, except when the lower boundary of the rooting depth is within the capillary rise of the ground water. Deep soils (such as soils M6 and F3) have a higher water-holding capacity, thus enabling the crops to bridge dry periods more easily than crops on shallow soils.

There is a clear precipitation surplus in autumn and winter, but no research has been done on the differential effect of leaching in the various soil/climatic regions. Dutch soils certainly undergo leaching, e.g. decalcification (see Van der Sluijs, 1970). There are podzols in the Netherlands (such as soil P3) but these are only developed in sands of late-Pleistocene age; loamy materials of a similar age, such as the loess from soil L1, carry Alfisols. Desalinization studies in Dutch polders flooded with sea water have indicated that on average 160 mm of precipitation is added to the ground water yearly (Verhoeven, 1953, p. 170 and 190).

In Soil Taxonomy there are five main classes of soil temperature regions: pergelic, cryic, frigid, mesic and thermic (SSS, 1975, p. 62 and 63). The Royal Netherlands Meteorological Institute measures soil temperatures at De Bilt, just east of Utrecht. The observation site is under grass, and most probably in a soil similar to soil P6. The observations started in 1914 at depths of 25 cm, 50 cm, 75 cm

Table 2. Climatic data of the Netherlands; monthly averages from 1930-61 at De Bilt (near Utrecht)

	J	F	M	A	M	J	J	A	S	0	N	D	Year
Sunshine (hours)	56.3	69.0	126.6	163.5	210.5	222.6	198.5	186.2	146.1	101.5	50.4	41.2	1572.4
Temperature (°C)	1.7	2.0	5.0	8.5	12.4	15.5	17.0	16.8	14.3	10.0	5.9	3.0	9.3
Maximum temperature below 0°C (days)	5	4	0	0	0	0	0	0	0	0	0	3	12
Minimum temperature below 0°C (days)	16	15	13	4	1	0	0	0	0	2	6	13	70
Maximum temperature 20°C or above (days)	0	0	0	2	9	15	21	21	11	1	0	0	80
Maximum temperature 25°C or above (days)	0	0	0	0	2	6	6	6	2	0	0	0	22
Maximum temperature 30°C or above (days)	0	0	0	0	0	1	1	1	0	0	0	0	3
Precipitation (mm)	68.0	52.2	44.6	48.8	51.5	58.0	76.8	88.0	71.2	72.2	70.0	63.4	764.7
Days with at least 1.0 mm precipitation	13	11	9	10	9	9	11	12	11	11	12	12	130
Evaporation (Eo) in mm, acc. to Penman	4	17	42	78	109	126	118	96	61	28	9	3	691

and 100 cm. The mean soil temperature at these depths are (1914-1974 period): 10.1 °C, 10.2 °C, 10.3 °C and 10.2 °C, respectively; the yearly fluctuation (means of ten-day periods) at 50 cm depth is 13 °C (Scharringa, 1976).

The mesic class of soil temperature region is defined as having a mean annual soil temperature of '8 °C or higher but lower than 15 °C, and the difference between mean summer and mean winter soil temperature is more than 5 °C at a depth of 50 cm' (SSS, 1975, p.63). Clearly the Dutch soil temperature region has to be called *mesic*.

Soil Taxonomy also defines soil moisture regimes (SSS, 1975, p. 51-57): the Dutch climate is such that nearly all soils satisfy the definition of the udic moisture regime, only the soils of the coastal marshes (such as soil M1) have a peraquic moisture regime. Because practically all our hydromorphic soils are artificially drained there are hardly any soils left with an aquic soil moisture regime.

Time

In roughly half of the area of the Netherlands the parent materials are of Holocene age (British equivalent: Flandrian); in the other half they are of Pleistocene age, and in less than 1% older formations outcrop (Fig. 2).

The parent materials of Holocene age are mineral and are of marine or fluviatile origin (mostly loamy and clayey) or are organic; the Pleistocene sediments are all mineral and are predominantly sandy (cover sands) with only a small part loamy (loess and glacial till). For detailed particle-size analysis see Table 1.

The boundary between the Holocene and Pleistocene sediments has been put at 8000 B.C., but more than three-quarters of the surficial Holocene sediments are less than a thousand years old. The upper part of the 100 000 ha of the drained lake-bottoms (see section 'The drained lakes and peat uplands district' in next chapter) was deposited in the mid-Holocene age, and sedimentation ceased between 3000 and 2000 B.C., depending on the site. However, these sediments were covered with peat shortly after sedimentation had stopped, and have only recently (Table 5) been revealed by peat cutting and subsequent drainage. If the very beginning of pedogenesis (soil-formation time zero) begins 'as soon as the sediments have been deposited and the influence of soil-forming factors, especially climatological and biological ones, i.e. aeration and vegetation, begins to be felt' (Pons & Zonneveld, 1965, p.13), then soil formation has only been under way in the last few hundred years since the reclamation of these shallow lakes. Only the marine sediments of 'The older lands' (see first section in next chapter) have been exposed for more than a thousand years, and surfaces of Roman age occur only locally, while older surfaces (late-Neolithic) are even rarer.

In the fluviatile district nearly all superficial deposits predate the construction of the artificial levees (between 1000 and 1300 A.D.). Few predate Roman times; but in those that do there is evidence of progressive soil formation, not only decalcification but also translocation of clay (De Bakker, 1965).

The upper part of the sediments in the Pleistocene district of the country consists of cover sands, an aeolian sediment from the late-Weichsel ice age (British equivalent: Devensian). For particle-size analysis see Table 1. The cover sands mostly date from 10 000 B.C.; in many places the superficial sands are somewhat younger, occasionally even from the boundary between Pleistocene and Holocene (8000

B.C.). These superficial sands have resulted from the erosion of the older cover sands, which they partly cover; locally, buried soils can be observed, consisting of a pale, somewhat bleached horizon, or sometimes an organic layer.

The sediments forming the coarse-textured hills in central Netherlands, from which soil P4 developed, are mostly Rhine sediments from the pre-Saale interglacial period, the Holsteinien (British equivalent: Wolstonian). The glacial till in the north of the country (e.g. the subsoil of soil P2) also dates from the Saale ice age. Due to erosion and solifluction in the tundra climate of the Weichselien the actual surface of these sediments is much younger than the age of the sediments themselves: it is also about 10 000 to 12 000 years old.

The loess in the south of the Netherlands is mostly as old as the older cover sands, namely dating from about 10 000 B.C.

To sum up, the surface of the Pleistocene district is mostly between 10 000 - 12 000 years old and that of the Holocene area mostly between 0 - 1000 years old. The surface of 'the older land' in the marine district is post-Roman, and in the fluviatile district the surface is only very rarely more than 1000 - 2000 years old.

Topography

Broadly speaking, the Netherlands slopes from the south-east to the northwest (Fig. 3), the highest point (321 m above sea level) being near the meeting-point of the boundaries of the Netherlands, Belgium and the Federal Republic of Germany; the lowest point is just north of Rotterdam, and is 6.6 m below sea level on the bottom of a reclaimed lake. To avoid confusion, no contour lines have been shown on the coastal dunes in Figure 3; generally they are between 10 and 30 m above sea level, the highest point being 56 m.

There are two irregularities in this general pattern: the hills in the centre of the country (highest point 103 m above sea level, north of Arnhem), and the 'holes' in the west of the country, scattered areas below the -2.5 m contour below sea level (lowest point, see preceding paragraph). The hills were formed by the Scandinavian ice sheet that pushed coarse river sediments into low hills in the Saale ice age; the holes are the reclaimed lakes, initiated by peat cutting (see the section about drained lakes in next chapter), the larger areas below sea level are the Zuyder Zee polders.

The general relief is mainly explained by the geological situation of the Netherlands; it forms a part of a large depression that is gradually sinking and being infilled with Quaternary sediments. This geosyncline is bordered in the south, in Belgium by the Cambro-Silurian Brabant Massif and the Devonian Ardennes, and in the south and east, in the Federal Republic of Germany, by the Rheinische Schiefergebirge (Eifel and Sauerland). The relief is partly explained by the occurrence of a horst-and-graben structure: the German part of the area below the + 30 m contour (Fig. 3) is part of the Lower Rhine Graben, and the elevation above 30 m west of the Meuse is part of the Peel Horst.

Far more important for pedology (and for agriculture!) are the small differences in elevation above the gently sloping ground-water table. The soil-forming factor 'relief' or 'topography' could better be replaced by 'depth to ground water' in the Netherlands.

Most subsoils of Dutch soils are moderately to rapidly permeable, and the water in the saturated zone can be characterized as *Grundwasser* (free ground water) and not as *Staunässe* (excess surface wetness). This is an important distinction made in the German and British systems of soil classification (Mückenhausen et al., 1977, p. 127; Avery, 1973).

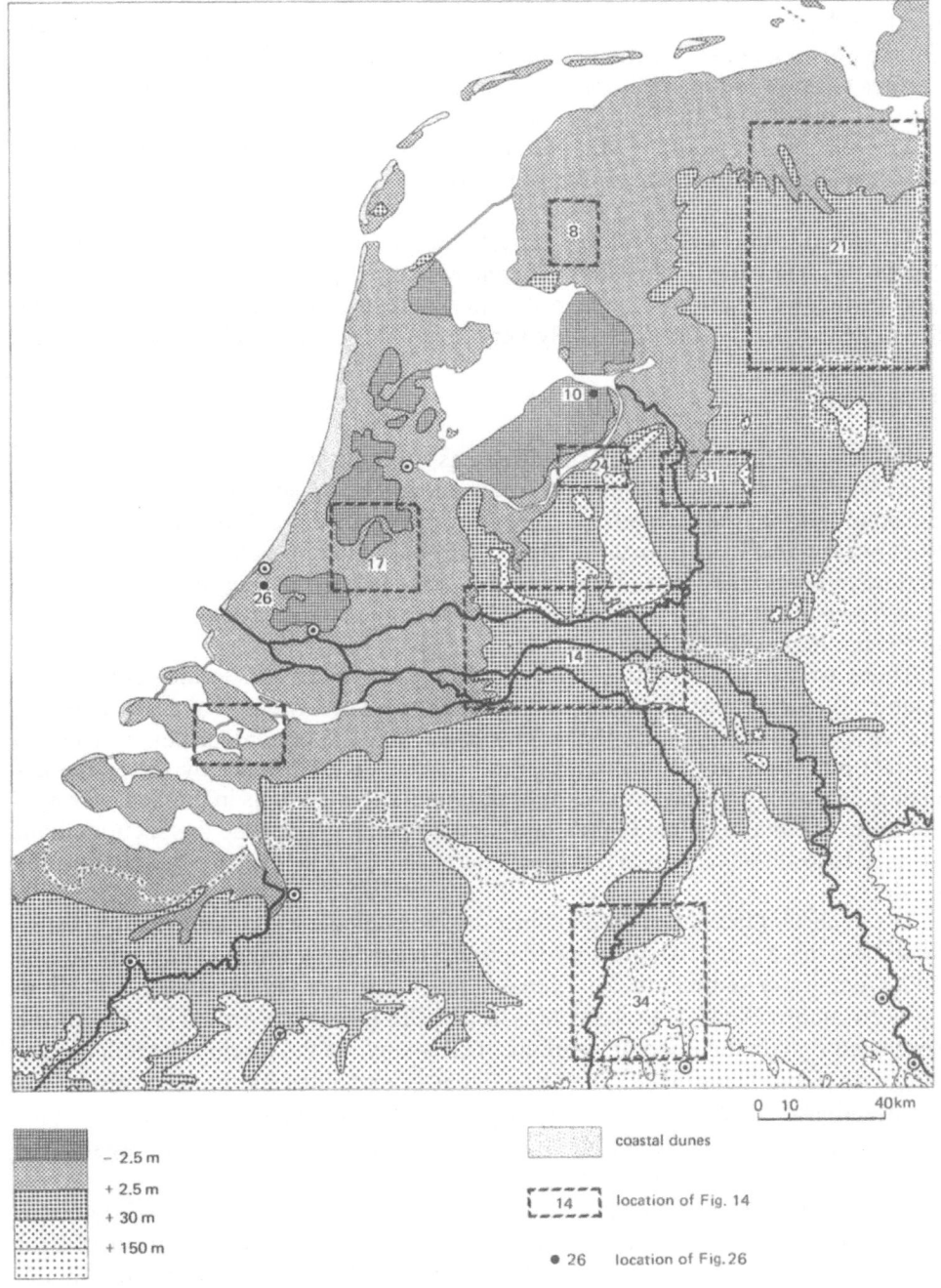

Fig. 3. Relief map (below or above sea level) of the Netherlands and the adjacent parts of Belgium and the Federal Republic of Germany, showing location of some text figures.

Table 3. Depth of the ground-water table in the Netherlands (% of the country)

Depth classes (cm below the surface)	0-20	20-40	40-70	70-100	100-140	140-200	> 200
In winter	24	25	27	11	7	3	3
In summer	0	1	14	19	28	27	11

The ground-water table fluctuates in the course of the year; even in artificially drained soils this fluctuation may be between 90 and 130 cm. The depth of the fluctuating ground water varies between just below ground level and a depth of several metres. This phenomenon is classified in seven 'water-table classes'. The definitions of these classes are based on the depths of the mean highest (MHW) and mean lowest water (MLW) tables (Van Heesen, 1970, p.268). For example water-table class III must have a MHW shallower than 40 cm and a MLW between 80 and 120 cm depth; water-table class VII must have a MHW lower than 80 cm depth and a MLW below 120 cm depth. On modern Dutch soil maps all mapping units are annotated according to soil and ground-water class, the latter is expressed in Roman numerals. This is true even for non-hydromorphic soils, such as soils A3, P3 and P4. All non-hydromorphic soils belong to class VII, but there are also hydromorphic soils that fit this class. Examples of this kind of well-drained soil are found on cover-sand ridges surrounded by raised bogs, which are now cut over, resulting in much lower water tables; they are even found in our coastal polders (Soil M6, p.82 has class VI).

These water-table classes were omitted from the soil maps in this book because the soil boundaries do not always coincide with the boundaries between the water-table classes; if I had inserted these lines the black-and-white maps would have been much too complicated.

Table 3 indicates depth of ground water in the Netherlands, these data are from a survey carried out from 1952-1955 (Visser, 1958). Ground-water levels were measured in auger holes or perforated pipes every two weeks (2000 wells) and four times a year (23 000 wells); the results were published in two series of 9 separate maps (summer and winter conditions). In the *Atlas of the Netherlands* these data were condensed in two maps, scale 1:600 000 (Atlas, 1963-1977, sheets VII-6 and VII-7). As a result of improvements in drainage since the 1950's the wetter classes of Table 3 will have decreased in favour of the intermediate classes.

Biotic factors

Animals. There is no department in the Dutch Soil Survey Institute studying soil biology. Soil microbiology is also beyond the scope of the Institute. In the Department of Soils of the Agricultural University of Wageningen, some research has been done on burrowing animals, such as earthworms and moles, which have obviously been active in the well-drained, calcareous soils F3 and LP3. Hoeksema (1953) called the process whereby these animals cause the gradual disappearance of soil lamination 'homogenization'.

Vegetation. In the Netherlands there is hardly any 'natural' vegetation, only some semi-natural and near-natural vegetations. The latter are only found in those areas called 'waste lands' in the introductory section to this chapter; these occupy 5% of the country.

The climatic climax vegetation on the mineral soils and on the eutrophic organic soils in the Netherlands must have been a forest, with alder, ash, beech, birch, elm, hornbeam, oak and willow in different combinations and with different undergrowth, depending on the site (rich or poor, calcareous or acid, waterlogged or well drained, or subject to flooding with fresh, brackish or saline water); on the oligotrophic raised bogs the natural vegetation was a treeless wilderness with peat mosses (*Sphagna* spp.) predominating.

However, many soils never had a forest vegetation before being used for agriculture; some never had a vegetation at all. All the coastal polders (one of the subregions of the marine district, see next chapter) that were enclosed and drained were reclaimed partly from saltings supporting a vegetation of salt-tolerant grasses and herbs, and partly from bare tidal mud flats. The Zuyder Zee polders were reclaimed from the sea bottom, and for only a few years supported a reed vegetation (*Phragmites communis*) that had been sown deliberately during reclamation to accelerate the ripening of the mud (Pons & Zonneveld, 1965, p.33). The hypothetical climax vegetation on these fertile, calcareous, well-drained polder soils would probably be an ash/elm forest.

Other soils, such as the base-rich well-drained soils of the natural levees in the fluviatile district (e.g. soil F3) carried some kind of ash/elm forest before reclamation, but there are no records or relics of such forests, for they disappeared more than a thousand years ago.

The Pleistocene district must have known the whole vegetational sequence from the end of the Weichselien until today. This sequence has been studied by palynologists by means of pollen analysis of organic layers. It is beyond the scope of this book to recapitulate the fluctuations in the pollen diagrams of pine and birch, the arrival of tree species such as hazel, oak and alder, and later of the beech and hornbeam and their fluctuations, and the arrival of cultivated plants and their accompanying weeds, indicating the start of man's agricultural activities.

Every treatise on vegetation as a soil-forming factor in the Pleistocene district in northwestern Europe has to take into account that vegetation has changed considerably since soil-formation time zero (roughly 10 000 to 12 000 years ago).

An important change in vegetation in the Pleistocene district is the gradual transformation of the forests on the poor sands into heathlands (cf. p.175). This started about 5000 years ago and it is assumed that Neolithic people induced this change by felling the trees for timber and for fuel and then burning and grazing the vegetation on soils that had a low potential for forest regeneration.

Man. In the densely populated Netherlands (in 1976: 409 inhabitants per km²) human influence is an important soil-forming factor. When using the old saying 'God created the earth, but the Dutch made their own country', most people point to the polders, particularly to the Zuyder Zee polders. However, on many pages of this book it is made clear that there are other soils and soil regions modified and reshaped by man. In the section 'Anthropogenic soils' in the next chapter, there is a discussion about the degree of human influence on Dutch soils and some examples are given of man-made soils, e.g. two plaggen soils.

Not only soil morphology but also soil fertility has been changed by man. As a result of the heavy application of fertilizers (cf. Fig. 30) the chemical fertility of

Dutch soils (even of soils that were originally poor or acid) is generally high. The fertility of the same kind of soil may differ depending on the kind of land use and the skill of the farmer or horticulturist; therefore no data about phosphate or potassium content are given in the 'Analytical data' accompanying the profile descriptions.

To give an idea about the intensive Dutch agriculture some statistical data will be given below. Although only 7% of the economically active people work in agriculture, the percentage of the total export of agricultural produce varies between 27% and 30%. There are 4964 thousand cows in the Netherlands, of which 2238 thousand are dairy cows (= 1.7 cows/ha grassland): the dairy cows produced 10 653 million kg of milk in 1976. Three-quarters of the consumption of nitrogen fertilizers (cf. Fig. 30) is used on grassland, which occupies 61% of the agricultural land. The increasing application of fertilizers has considerably increased crop yields, e.g. the average yield of rye increased from 1200 kg/ha in 1870 to 3500 kg/ha in 1975 (cf. Fig. 30).

In 1975 the average yield of winter wheat was 5100 kg/ha, of potatoes 32 900 kg/ha and of sugar beet 43 500 kg/ha.

The growers of vegetables and flowers are organized in auction societies. All their produce has to be marketed in central auction buildings (cf. Fig. 26). In 1974 1441 million kg of vegetables were auctioned, worth Dfl 1200 million, and the 1974 turnover of all flower auctions was Dfl 954 million.

Soil geographic districts

Marine district

The area of soils, developed from Holocene marine parent material (Fig.2), occupies about 965 000 ha, or a good quarter of the total area of the Netherlands. It can be subdivided into four regions: the bottomlands in the drained lakes; the coastal polders enclosed after 1200 A.D.; older reclaimed land situated further inland; and the recent Zuyder Zee polders. The main differences between these regions are indicated in Table 4.

Table 4. Differences between the four regions within the marine district of the Netherlands

	Drained lakes	Coastal polders	Old land	Zuyder Zee polders
Origin	shallow man-made lakes	coastal marshes outside dikes	natural forelands, bordering peat and Pleistocene	shallow, wide marine lake
Age of the upper part of the parent material	3000-1500 B.C.	1100 A.D.-present	1500 B.C.-1100 A.D.	1800 A.D.-present
Period of enclosure	1500-1950 A.D.	1200 A.D.-present	before 1200 A.D.	1930-present
Elevation	4-6 m below sea level	− 0.5 − + 1.5 m	− 1 − + 0.5 m	4-6 m below sea level
Soils — carbonates and acidity	calcareous and non-calcareous, non-acid and acid to strongly acid	calcareous	non-calcareous and shallowly non-calcareous, non-acid	calcareous
Soils — texture topsoil	peaty, medium and fine	coarse, medium and fine	medium and fine	coarse, medium and fine
Soils — texture subsoil	medium and fine	coarse, medium and fine	medium and fine or peat and Pleistocene	medium, often organic-rich
Land use	arable and grass	arable	grass and arable	arable
Fields	regular, medium-sized	regular, medium-sized	irregular, small-sized	very regular, large-sized

The bottomlands in the drained lakes. This region is a part of the 'Drained lakes and peat uplands district'. It occupies about 100 000 ha (Table 5, p.32) and practically coincides with the area below the 2.5 m below sea level contour in the west of the country (Fig. 3). The uppermost of these sediments (which were deposited in the mid-Holocene age at a level about 4 m below present sea level) are called 'old marine clay' (Edelman, 1950). Stratigraphically these correspond with the 'Blue Clay' of Romney Marsh in southern England (Green, 1968, p.9 and 14) and with the 'Fen or Buttery Clay' in the English Fenlands of East Anglia (Seale, 1975, p.13, 20 and 21).

The geology, history and soil conditions of this region are discussed later in this chapter (p.30-34).

The coastal polders. In 1948 Van Veen estimated that 381 686 ha of tidal marsh had been reclaimed in the Netherlands. Taking into account the land empoldered since 1948 the total area of coastal polders today is approximately 400 000 ha (Fig.6). This figure does not include the drained lakes nor the Zuyder Zee polders, nor the older land. The fluctuations in the rate at which land was embanked are not easy to explain. Physiography, technical ability, land and crop prices, and politics were important. The period of the greatest activity in empoldering was 1600-1625 A.D. when 32 000 ha were reclaimed: this period includes the Twelve Years' Truce (1609-1621) in the war with Spain (cf. also Table 5).

A detailed history of the empoldering of an area in the south-west of the Netherlands may be pierced together from the innings data given in Figure 7. In this area, only narrow estuarine inlets bordered by natural levees or by raised bogs further inland probably existed in Roman times. The area must have resembled the 'Drained lakes and peat uplands district' (p. 30-34) before human settlement. The post-Roman marine transgression easily breached the levees, which had been undermined by peat-cutting, widened the estuaries, partly eroded the peat land and covered the whole area with fresh marine sediments. Further south, buried Roman settlements have been discovered, both on the natural levees and on the raised bogs.

In early mediaeval times the area was characterized by many shallow tidal gullies interspersed with shoals and bare sand flats that were submerged during high tide; on the higher sites the partly vegetated mudbanks were subject to occasional flooding only. Along the Pleistocene coast in the southeast (just outside the area depicted in Figure 7) the landscape elements were the same: coastal marshes grading into lower-lying sand flats. Enclosure started on the saltings fringing the higher mainland and on the mudbanks to the west; these areas are shaded dark on the map. Gradually the single-polder islands coalesced into larger multi-polder islands, or became linked to the mainland; the latter is the case with the former island shown in the aerial photograph (Fig.4).

As a result of the reclamation the tidal gullies became fewer and narrower; subsequent scouring deepened them. If this occurred where there was little or no foreland, the dikes became vulnerable to breaching. This happened directly west of the village on the photograph, where in 1682 the dike broke and 7 ha of land were lost. The narrow part of the estuary southwest of the enclosure dam built in 1964 is today the deepest spot (37 m below sea level) in the area shown in Figure 7. In the Spanish-Dutch War, the estuary at this very spot was shallow enough to enable the Spanish commander Mondragon to march his troops through shoals and over sand flats during ebb tide to the island in the west to besiege and to conquer the town of Zierikzee in 1575.

The coastal polders, called New Land by Edelman (1950), are situated roughly between 0.5 m below and 1.5 m above sea level, the youngest polders being the highest because of the ever-rising sea level.

Most of the boundaries between the fields in the coastal polders are 'wet fences' (ditches, acting as open drains) and, as can be seen from the aerial photograph (Fig.4), these are generally regular and the fields are reasonably large. Irregularities in the field boundaries are due either to irregularities in the polder boundaries (the sea walls, which are always grass-covered and used as pasture for sheep), or to the meanderings of former tidal creeks, which now remain as sinuous ditches.

In the area shown in the photo the size of the farms ranges between 40-80ha. The

Fig. 4. Part of the marine district in southwestern Netherlands. In the west the village Sint Philipsland (a) (population 2000, including the rural population) can be seen on the former island of the same name. In 1884 this island was connected to the mainland by a dam (b), the isthmian polder (c) north of this dam was enclosed in 1907. Today the situation has been completely changed by the construction of the Scheldt-Rhine Canal, which cuts right across the isthmus. In the saltings (d) the tidal creeks with narrow natural levees form a dendritic pattern; the bigger creeks are still recognizable in the polders as curved ditches (e). Same area as Figure 5.

settlement pattern is typical for the coastal polders: a small village in the oldest polder and scattered farmsteads in the surrounding polders.

Almost 90% of these polders are indicated on the 1:200 000 Soil Map of the Netherlands (Stichting voor Bodemkartering, 1961, sheet 11) as 'calcareous and slightly calcareous young sea clay soils', and are subdivided according to the texture profile (cf. Green, 1968, p.64) and the texture of the topsoil.

Figure 5 gives a detailed picture of soil conditions in such polders and also of the tidal flats outside the dikes. The aerial photograph (Fig.4) was taken during ebb tide. Only the deep gullies are water-filled (black on the photo, white on the soil map). The soils of the sand flats (the lowest parts of the unsubmerged area) are called Psammaquents; they are flooded twice daily. The higher parts support a dense vegetation of many different halophytic species; in the zone between 1m below and just above sea level cord grass (*Spartina x townsendii*) is the predominant plant species. The soils of this area becomes more clayey further inshore and are flooded less frequently than the sand flats; they are called Hydraquents, and soil M1 is an example of such a soil. Within the area enclosed by the dikes the mapping units are delineated according to differences in the texture of the topsoil and the depth to the sandy subsoil. Deep soils such as soil M6 (called Typic Fluvaquent) are encountered. Shallow Fluvaquents (which I propose to call Psammic) such as soil M4, are also found. Locally there are soils such as soil M5, where former sandbars outcrop. These are not present on this soil map.

The older land. This region covers an area of about 300 000 ha, and is situated roughly between 1 m below and 0.5 m above sea level (Fig.3). Inland the sediments thin out over peat, fluviatile materials or Pleistocene sands (Fig.2); seawards they

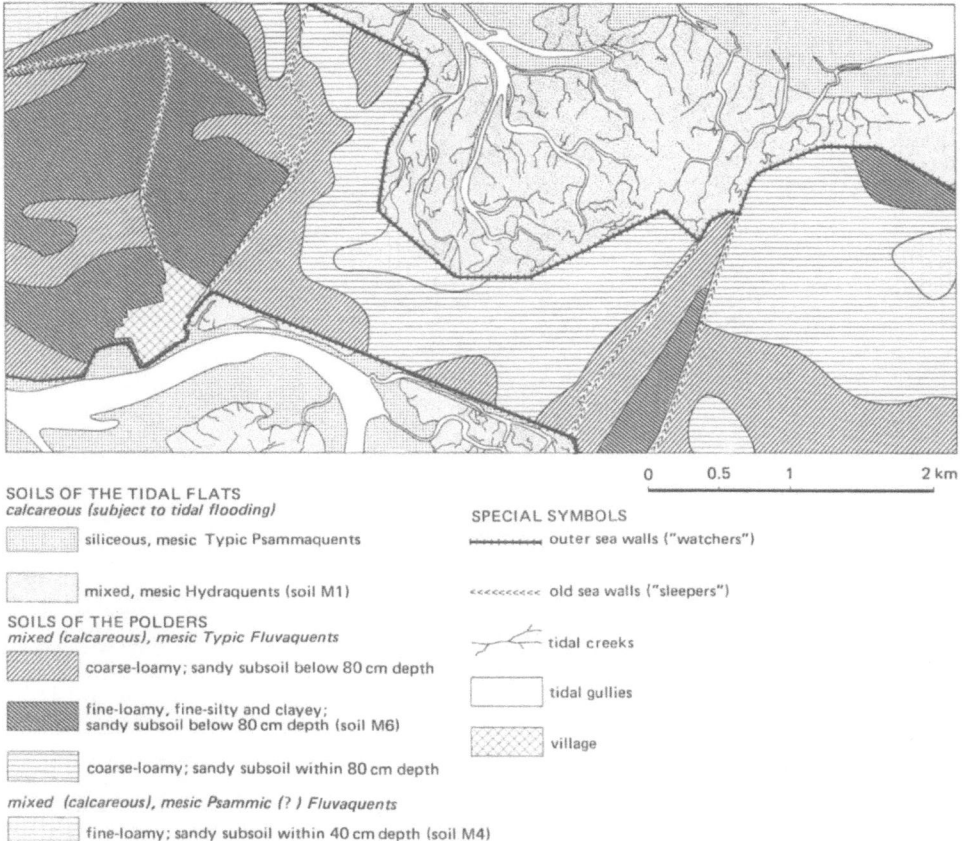

SOILS OF THE TIDAL FLATS
calcareous (subject to tidal flooding)

⬚ siliceous, mesic Typic Psammaquents

⬚ mixed, mesic Hydraquents (soil M1)

SOILS OF THE POLDERS
mixed (calcareous), mesic Typic Fluvaquents

▨ coarse-loamy; sandy subsoil below 80 cm depth

▨ fine-loamy, fine-silty and clayey;
sandy subsoil below 80 cm depth (soil M6)

⬚ coarse-loamy; sandy subsoil within 80 cm depth

mixed (calcareous), mesic Psammic (?) Fluvaquents

⬚ fine-loamy; sandy subsoil within 40 cm depth (soil M4)

SPECIAL SYMBOLS

⊢⊢⊢⊢⊢ outer sea walls ("watchers")

<<<<<<<<< old sea walls ("sleepers")

tidal creeks

⬚ tidal gullies

▨ village

Fig. 5. Simplified soil map of an area with coastal polders. Same area as Figure 4; for location see Figure 7.

are covered (where they have not been eroded) with the more recent sediments of the coastal polders.

In the southwest most of the older land is eroded. Scattered remnants occur only on some of the Zeeland islands, surrounded by the younger coastal polders. Some of these even abut on to the Pleistocene district, the marine sediment was brought in by the big tidal inlet scoured after the disastrous St-Elisabeth's Flood that occurred on 19 November 1421; this is now the estuary that carries the water of both Rhine and Meuse to the North Sea.

On the other hand, the belt of coastal polders in the north is narrow, and most of the marine district there is composed of older land. In many places settlement pre-dates empoldering: people lived on the higher land (natural levees of tidal creeks or on coastal barriers along the coast of that time). These sites were raised into *terps* (dwelling mounds) during periods of marine transgression; some of these *terps* date from before Roman times, some from as late as 800-900 A.D., but most of them were built in between these times. Pliny the Elder (in his *Naturalis Historia*) was probably referring to the inhabitants of these *terps* when he wrote of the *misera gens*, living on man-made elevations in an area subject to tidal flooding.

The differences between the older land and the coastal polders are indicated in

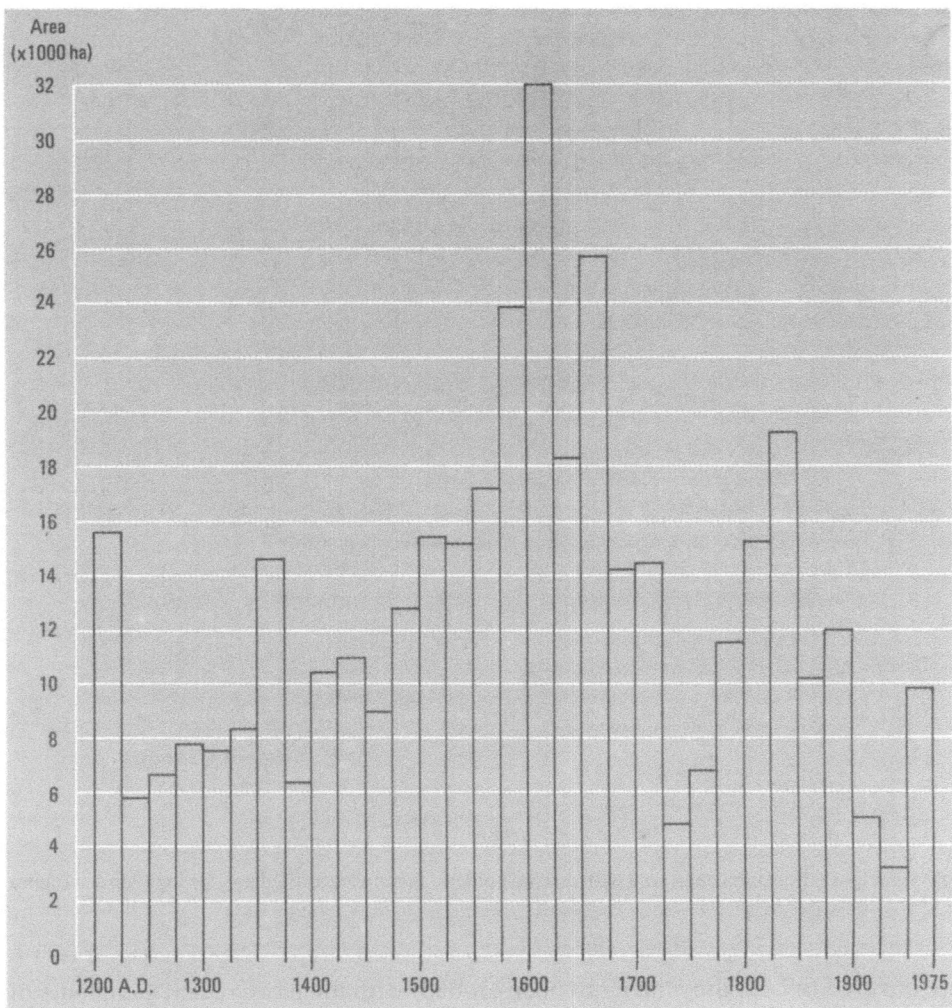

Fig. 6. Bar graph of the enclosure history of the coastal polders; each bar indicates the area (in ha) enclosed in a period of 25 years.

Table 4. The enclosure date, before or after 1200 A.D., is not completely arbitrary. In fact few dikes were built before this date, and the original outer dikes of the older land are often the original dikes. They were built to protect existing land against rising sea level rather than to conquer new land. The history of enclosure of coastal forelands in the Netherlands really began with the offensive dike-building after about 1200 A.D. A Dutch archaeologist who has investigated many *terps*, states that empoldering before the 12th century was insignificant and that before 1000 A.D. the absence of a central authority (emperor, count, bishop or monasteries) precluded extensive embankment schemes (Van Giffen, 1964, p.281).

Figure 8 depicts the surface geology across the transition from the thick marine deposits of the older land in the north-west via the peat soils with thin marine cover to the outcropping Pleistocene in the southeast. Both the marine deposits and the peat gradually taper out against the Pleistocene, as is shown in Figure 9, a sketch-

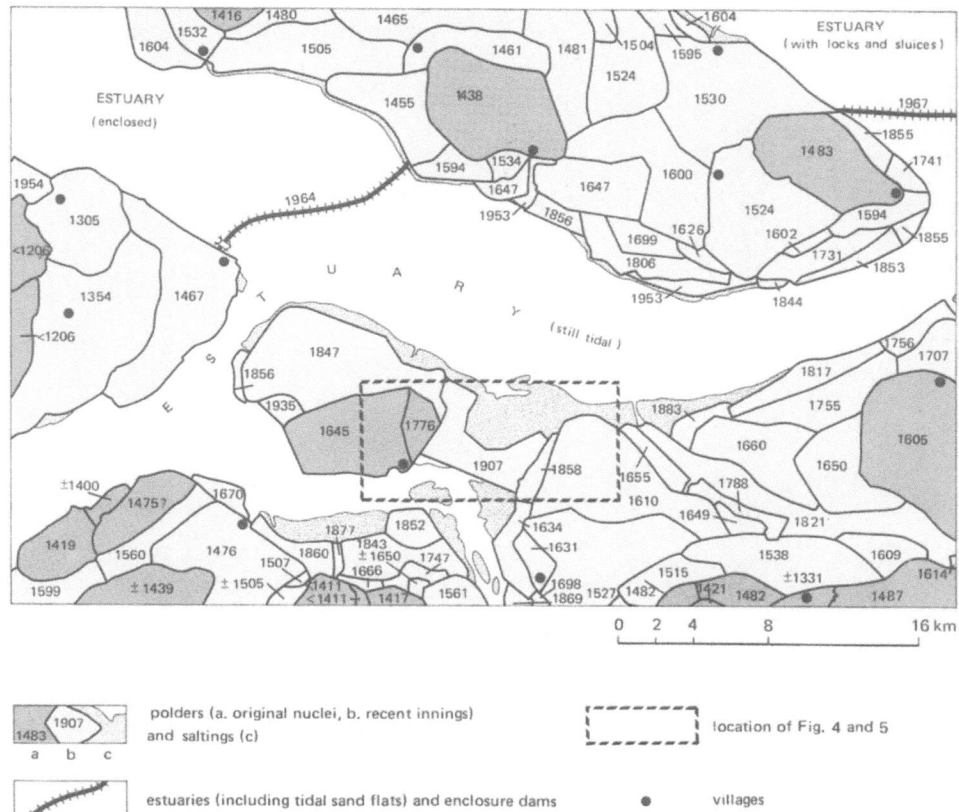

Fig. 7. Innings map of an area with coastal polders in the northeast of Zeeland Province, with enclosure dates. The scattered original nuclei accentuate the fact that this 'Old Zealand' was indeed formerly a 'sea-land'. The old polders in the southeast are embanked against the mainland (Pleistocene). For location see Figure 3.

section over a distance of about 16 km.

Marine materials deposited far in the hinterland are always fine-textured and non-calcareous, e.g. the topsoil of soil M7 and the upper 90 cm of soil M3. Soil M7 was sampled on the transition to the Pleistocene, and soil M3 was sampled in an area just beyond the northwest corner of Figure 8.

Further north there is a system of coastal barriers, forming the boundary between the older land and the coastal polders, the first of which was enclosed in 1505. These coastal barriers are the highest parts of the older land and have many *terps*, (Lambert, 1971, p.78) some of which were first settled in pre-Roman times. The soils on these coastal barriers are mostly sandy loam. Soil A5 comes from this area, but is described in the section on Anthropogenic soils, because it has been artificially raised.

The Zuyder Zee polders. The Zuyder Zee is, or rather was, a shallow marine lake connected with the North Sea via the Wadden Sea through the tidal inlets between the Frisian Islands. It is in the middle of the country and before the polders were reclaimed its surface area was 3700 km².

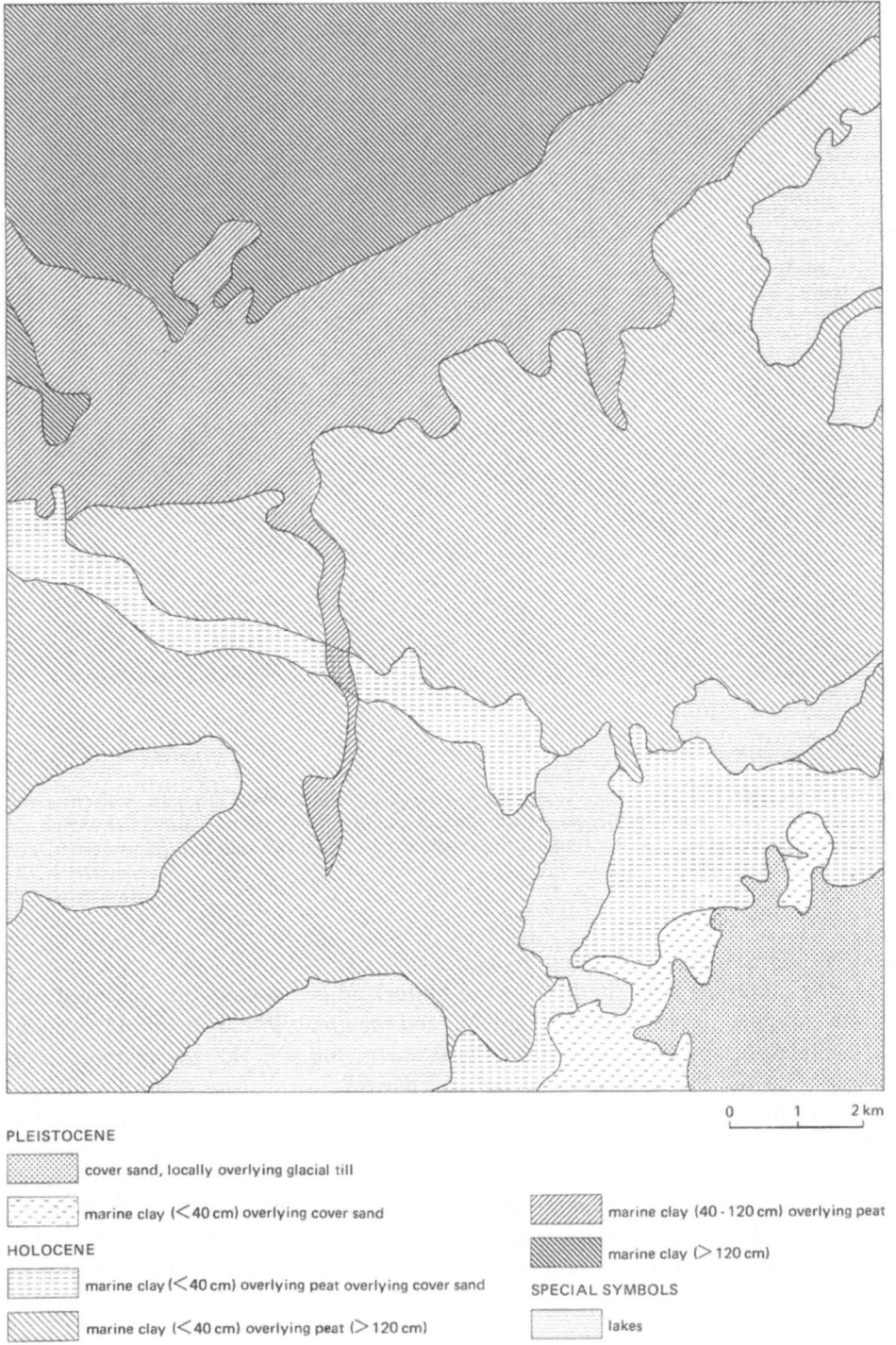

0   1   2 km

PLEISTOCENE

cover sand, locally overlying glacial till

marine clay (<40 cm) overlying cover sand

HOLOCENE

marine clay (<40 cm) overlying peat overlying cover sand

marine clay (<40 cm) overlying peat (>120 cm)

marine clay (40 - 120 cm) overlying peat

marine clay (>120 cm)

SPECIAL SYMBOLS

lakes

Fig. 8. Surface geology and parent materials in an area of older lands in the marine district. The Holocene marine sediment thins southwards over the peat. The rise of the sea level was more rapid than the upward growth of the *Sphagnum* peat, thus drowning the raised bog. Still further south this post-Roman sediment peters out against the outcropping Pleistocene. For location see Figure 3.

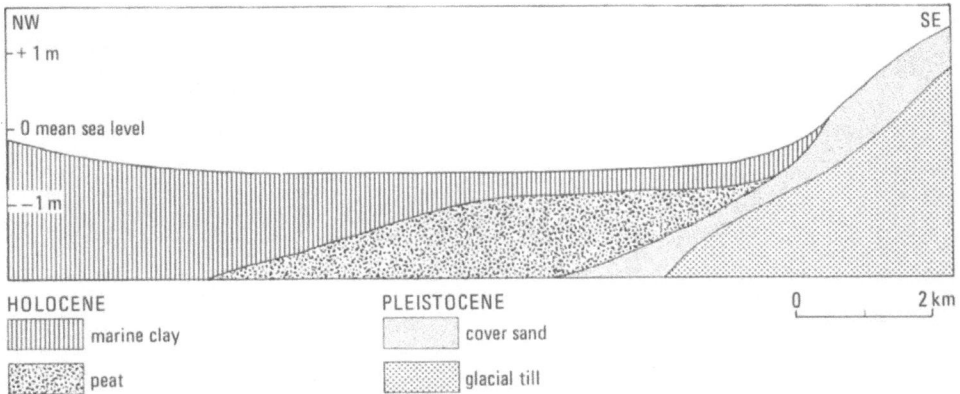

Fig. 9. Sketch-section through marine deposits overlying peat, both wedging out against Pleistocene sediments. Differential shrinkage explains the difference in elevation between the surfaces of the deep marine sediment and the shallow sediment overlying the peat. The section is aligned from the northwest to the southeast of Figure 8.

About 2000 B.C. this area was part of the large coastal belt of peat overlying Pleistocene sands and, further west, overlying older Holocene marine deposits. After that time the marine influence increased and the easily erodable peat gradually disappeared, to be replaced by marine sediments. This erosional and depositional history has been studied in detail before and during the embankment and reclamation of the polders. To quote Wiggers's summary: 'Roman writers described the presence of a large lake, called Lake Flevo, in the interior of the Netherlands. This lake was situated in the central part of the former Zuyder Zee ... In 755 this lake was called Almere and in 1340 Zuyder Zee. Meanwhile it had changed into an enlarged bay, which became more and more brackish. In 1932 the enclosure of the Zuyder Zee created a fresh-water lake, called Lake Yssel' (Wiggers, 1955, p.201). The lower part of the sediments have a high organic-matter content, caused by the incorporation of peat eroded from the banks of the continuously expanding lake. The upper layers contain less organic matter, which is more normal for Dutch marine sediments. The organic-matter profile of soil M2 illustrates this clearly.

Four polders have been reclaimed to date inside Lake Yssel; they practically correspond with the area lying more than 2.5 m below sea level. The first polder was embanked in 1930 (20 000 ha), the second in 1942 (48 000 ha). The third and the fourth polders together form a double polder, the northern half of which was embanked in 1957 (54 000 ha) and the southern half in 1968 (43 000 ha). The enclosure of a fifth (the last Zuyder Zee polder) is still in progress in the west. After it has been completed the remaining Lake Yssel will have less than a third of the original surface area of the former Zuyder Zee: about 1200 km².

The polders are enclosed and reclaimed by Governmental Authorities, who are responsible not only for constructing dikes and pumping stations, but also for digging the ditches and drainage channels, tile-draining the fields, and constructing roads and farmsteads. Afterwards the farms are leased to individual tenants.

The fields in these polders are rectangular. The development of agricultural machinery caused the planners to enlarge the fields in successive polders, from 20 ha, 24 ha and 30 ha in the first three polders respectively, to 60 ha and even 90 ha in the fourth polder. As appears from Figure 10 the dimensions of the fields in the

Fig. 10. Northern part of Oostelijk Flevoland (54 000 ha), the second youngest (1957) polder in the former Zuyder Zee. Farmsteads are situated along roads, the fields are tile-drained into ditches (a) draining into large drains (b) which in turn lead into the main drainage channels (c) which convey the water to the pumping stations. At (d) is one of the three pumping stations of this polder, this one discharges into Lake Ketel (e), a part of Lake Yssel (water surface 20-40 cm below sea level). A lock (f) separates the two pumping districts of this polder (pumping levels 6.20 m and 5.20 m below sea level). For location see Figure 3.

third polder are 300 x 1000 m (30 ha); most of the farms in the photograph have one field, but some have two. Soil M2 was sampled in the centre of this polder and its features are typical for the southeastern part of the basin of the former Zuyder Zee: fine-textured, strongly cracked and overlying Pleistocene sand.

Fluviatile district

Holocene sediments, associated with the Rhine and Meuse (Fig. 2) occupy about 270 000 ha, or nearly 8% of the total area of the country. When it enters the Netherlands, north of Antwerp, the river Scheldt is already a tidal estuary and its sediments are classified as marine deposits.

At the border with Germany, a few km southeast of Arnhem the Rhine is about 15 m above sea level. It bifurcates twice shortly after entering the Netherlands; the major branch is the Waal (the *Vahalis* of the Romans), which is the shipping lane between Rotterdam and the Ruhr. The two smaller branches are called the Yssel (which flows north to Lake Yssel) and the Lower Rhine which flows west past Arnhem and Wageningen. The Meuse leaves Belgium just north of Luik (Liège) at about 50 m above sea level. Both river plains converge at about 6 m above sea level; the Meuse and the Waal-branch of the Rhine actually meet at about 4 m above sea level (Fig. 14). In 1904 a shipping canal with locks was constructed at this point. The fluvial sediments pass beneath the marine sediments at 0.5 m above sea level.

The transition from floodplain to the adjacent higher ground formed from Pleistocene deposits is generally gradual, because these older sediments dip very gradually under the fine-textured sediments of the backswamps. In a few places the river plain is bordered abruptly by 20-30 m high bluffs, caused by river erosion of the southern slopes of the ice-pushed hills.

The Pleistocene outcrops shown in the middle of Figure 14 are partly buried, fossil river dunes bordering buried channels of the braided river systems of the Pleistocene Rhine and Meuse. These have a westward slope of about 30 cm/km and near Nijmegen disappear beneath Holocene fluviatile sediments, which have a lower seaward gradient of 10-15 cm/km.

The floodplain has three major elements: forelands, natural levees and back-swamps (Fig. 14).

The forelands (called washes or washland by Edelman, 1950, p. 54 and 55) lie outside the artificial levees or river walls, and are subject to flooding. Consequently there are no villages or farmsteads on the forelands, a conspicuous feature on the aerial photograph (Fig. 11). On the forelands land use is limited to pasture and meadow. The hay is protected from the rare, and mostly low, summer floods, by low banks, called summer dikes. During periods of high discharge by the rivers the forelands are flooded (Fig. 13). Flooding across the forelands enables high flood levels to be contained in the rivers. The main river banks are much higher than the summer dikes, and must be able to stem higher floods, which occur mainly in winter; hence the name winter dikes. The area lying landwards of this dike is only flooded when the dike is breached. This happened frequently in the past, as is witnessed by the many scour holes which remain (Fig. 11), in the southern dike these scour holes date from 1811, 1726, 1781 and 1861 (from left to right respectively). During the last winter of the Second World War the riverine area was no-man's-land after the Battle of Arnhem. On 3 December 1944 the southern dike of the Lower Rhine west of Arnhem was blown up by the Germans and practically the whole area between the Waal, the Amsterdam-Rhine Canal and the Lower Rhine was flooded (Fig. 13). The previous time that this area was inundated was in 1855, when the southern dike of the Lower Rhine was breached on four places at a time of high floods.

Brick making is the only industry on the forelands; the chimneys of the brick kilns are a typical feature of the skyline. The buildings are situated on elevated sites surrounded by partly excavated areas (Fig. 12).

Actual and fossil river courses in the Netherlands are accompanied by ridges that are 1 m higher than the backswamps. These ridges range from several hundred meters to 2 km wide, and are called natural levees. All the villages (Fig. 14) and old roads are situated above the floodplain on these levees. Most villages pre-date embankment (roughly before 1200-1400 A.D.). The centre of many villages is slightly elevated and has deep, dark soils where medieval or even Roman artefacts may be found. During the restoration of the war-damaged church of Elst, a village southwest of Arnhem, remnants of a Gallo-Roman temple were found.

The soils on the levees, called river ridge soils by Edelman (1950, p. 40), are characterized by a medium-textured, well structured upper part of the solum overlying a coarse-textured subsoil at varying depth, resulting in a well-drained soil. The relative elevation of these soils enables them to be used mainly for arable land and orchards. The soils on the levees of the Rhine and its lower branches are mostly calcareous at a shallow depth. Locally there are remnants of levees that are 3000-5000 years old and which have survived a rejuvenation of the meander belt. Soils at such sites show progressive stages of development; not only decalcification but also evidence of clay translocation (De Bakker, 1965). By comparison, the

Fig. 11. Part of the fluviatile district near the junction of the Amsterdam-Rhine Canal with the river Waal, east of Tiel (a) (regional market town, population 23 000). Urban areas are restricted to the natural levees, both forelands and backswamps are practically devoid of inhabitation. The isolated water-holes have various origins: those bordering the dikes are scour holes (b) (scars of dike breaches); the narrow ones on the forelands are a kind of bayous (c) (abandoned water courses); the others on the forelands are borrow pits (d) from the brick kilns; and those inside the dikes are very deep holes where sand has been sucked to build the ramp for the new bridge. Same area as Figure 12.

levees of the Meuse are non-calcareous throughout. This difference in lime content between the levees of the Rhine and the Meuse is inexplicable.

The soils of the levees grade gradually into the backswamps, the third element in the Rhine-Meuse floodplain. Edelman (1950, p.48) spoke of 'basins' referring to their low-lying situation between the surrounding levees (Fig. 14). The same situation has been described in the Trent Valley in England (Bridges, 1973, p.248) and in the Mississippi Valley (Fowlkes et al., p. 9-12).

Until the end of the Second World War the landscape of the backswamps was characterized by widespread grasslands, lacking farm buildings and with only some gravel and unsurfaced roads leading to meadows used for extensive grazing. Trees were limited to scattered willow-coppices, some surrounding duck decoys. The soils in these areas are fine-textured, non-calcareous and sometimes have peat at shallow depth. Until fairly recently these soils were badly drained, because ground-water levels are high and drainage conditions poor.

In the last twenty years many of these areas have been reclaimed, including the backswamp south of the Waal (Fig. 11). Many new metalled roads, lined with

| 0 | 0.5 | 1 | 2 km |

SOILS OF THE FORELANDS
calcareous, not classified

undisturbed

excavated for brick making

SOILS OF THE EMBANKED AREA
mixed, mesic Fluventic Eutrochrepts (calcareous)

coarse-loamy, over coarse loamy or over sandy

fine-loamy or silty, over coarse-loamy or
over sandy (soil F3)

mixed, mesic Typic Fluvaquents (calcareous)

fine-loamy or fine-silty, over fine-silty or
over clayey

illitic, mesic Typic Fluvaquents (non-calcareous)

deeply fine-clayey (soil F1)

SPECIAL SYMBOLS

main artificial levee (winter dike)

small artificial levee (summer dike)

urban and industrial areas

water and marshy areas

brick kilns

railway

four-lane highway (partly under construction)

Fig. 12. Simplified soil map of a part of the fluviatile district. Same area as Figure 11; for location see Figure 14.

poplars, now cross the backswamps. Originally most villages were connected only by roads on top of the winding dikes. Comparison between Figures 11 and 13 shows a new four-lane highway north of Tiel connected by a road and a bridge (in the photograph still under construction) to the villages south of the river. All the villages on the southern levee of the Waal (Fig. 14) have now been linked to the city of Nijmegen by a new secondary road. Modern dairy farms have been built in the

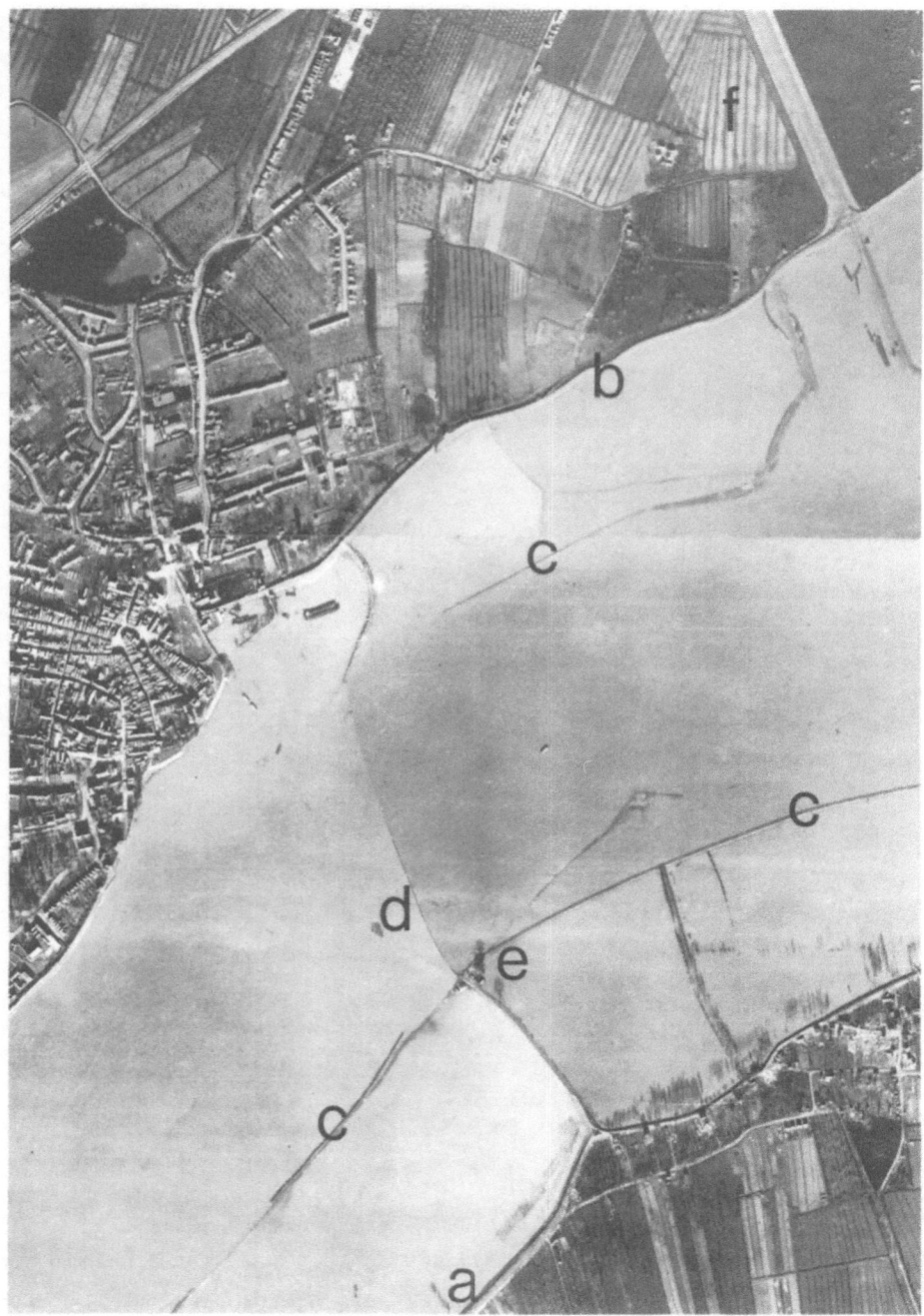

Fig. 13. Submerged flood plain of river Waal in the winter of 1945 between the southern·(a) and the northern (b) winter dike, with partly submerged summer dikes (c). The cluster of trees in the middle of the water (d) marks the northern end of the raised road to the ferry, on the crossing with the summer dike (e) the isolated ferry-house. On the north bank (cf. Figure 11) the highway has not been built but the canal (f) is under construction. Luckily the eastern dike of the canal had been completed: it formed the western boundary of the inundated area caused by the blowing up of the Lower-Rhine dike west of Arnhem (3 Dec. 1944). Same area as northwestern part of Figure 11.

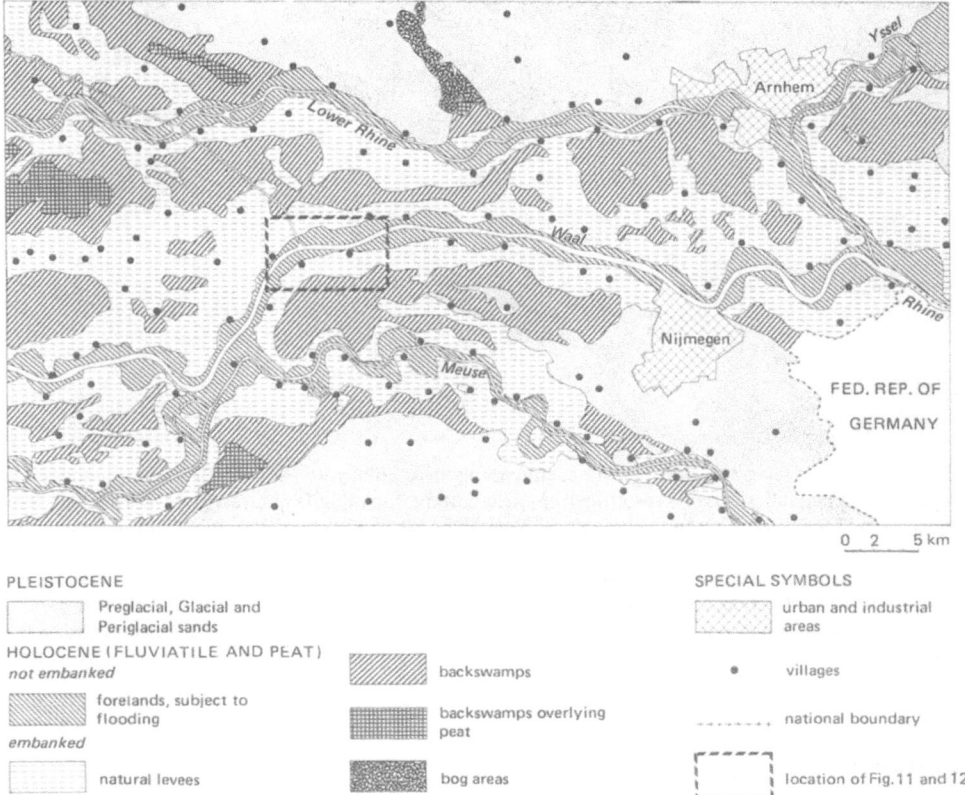

PLEISTOCENE

Preglacial, Glacial and Periglacial sands

HOLOCENE (FLUVIATILE AND PEAT)
not embanked

forelands, subject to flooding

embanked

natural levees

backswamps

backswamps overlying peat

bog areas

SPECIAL SYMBOLS

urban and industrial areas

• villages

- - - - - - - national boundary

location of Fig. 11 and 12

Fig. 14. The main Holocene geomorphological elements in the fluviatile district. For location see Figure 3.

former uninhabited backswamps. With the reallocation schemes drainage conditions have been improved by digging new main drains and installing modern pumping stations to lower water levels, and by better maintenance of the field ditches.

Three profiles have been selected to illustrate the soil conditions in the fluviatile district. Soil F1 is from a backswamp and the other two are from a Rhine levee.

The high clay content of soil F1 illustrates the calm conditions during sedimentation in this former backwater area. The presence of the buried A1 horizons, one dating from the Bronze Age, the other from Roman times, demonstrates that this site has been free from erosion by river meanders for over 3000 years.

Soils on levees have a coarse sand subsoil; F2 is an example of a shallow soil and F3 of a deep soil (coarse sand 40 cm and 130 cm deep respectively). Shallow soils are rare and are not mappable at scales of 1 : 10 000 and smaller, so they are included as impurities within other mapping units. In dry summers the shallow soils are clearly recognizable by differences in the colour of sugar-beet leaves.

On the soil map (Fig. 12) soil F1 fits the description of mapping unit 6, and soil F3 of mapping unit 4. Soils with sand at a very shallow depth, such as soil F2, are rare in this mapping unit.

Fig. 15. Drained lakes (a) with marine sediments as lake bottom (bottomland), un-cut peatland (b) (upland) and two lakes (c), one resulting from peat cutting, the other is natural (in the northwest). The villages and most roads are on the uplands alongside the fen streams which originally drained the now vanished raised bogs. The upland soils are partly mineral (cf. Figure 16), in so far as they are organic (soil LP1) after grassland, horticulture is important; mostly greenhouse flowers such as carnations, roses and chrysanthemums. Outside this picture, near the village of Boskoop, ornamental shrubs are an important crop. Arable land is dominant on the calcareous lake-bottom soils (soil LP3); grassland on the other soils – some of which are acid sulphate soils (soil LP2). In the western part roddons or fossil tidal creeks (d) are visible. Same area as Figure 16.

Drained lakes and peat uplands district

This district lies northeast of the line linking Rotterdam and The Hague and extends about 40 km north of Amsterdam. On the map of parent material (Fig. 2) it is characterized by a complex pattern of peat, loamy marine and fluviatile material, and on the relief map (Fig. 3) by scattered areas below the 2.5 m below sea level contour. To explain this complex pattern, it is necessary to discuss the geological and human history of this area, which is well known.

Approximately 4000-5000 years ago the western part of the Netherlands was a coastal lagoon, lying about 4 m below the present sea level. There were many tidal creeks, some of them forming part of the network of the lower distributaries of the Rhine and the Meuse. It is known that Neolithic people lived on the natural levees in this area. From excavations of their dwelling places it appears that they fished for sturgeon and even porpoises in the rivers and tidal creeks, and collected mussels on the tidal flats. This area was protected by low coastal barriers.

At the end of the third millennium B.C. during a regression of the sea, the gaps in these barriers narrowed gradually and only a few continued to function as river outlets. The marshy lagoon became desalinized by rain and river water and peat began to accumulate over the former tidal flats, first as a reed swamp (on the lowest places Phragmites peat is present). As the peat got thicker the fen became solely dependent upon rain water and it changed into a raised bog with peat mainly derived from Sphagnum moss. Along the rivers the land was occasionally flooded, and the eutrophic situation was maintained only in strips bordering the rivers.

When the Romans arrived in this area they found small rivers with narrow natural levees covered with deciduous forest, mainly oak. Behind the levees the

SOILS OF THE UPLANDS
mixed (non-acid), mesic Typic Fluvaquents

       deeply fine-silty

illitic (non-acid), mesic Thapto-histic Fluvaquents

       clayey over histic

euic, mesic Typic Medihemists

       mucky clay loam over peat (soil LP1)

SOILS OF THE DRAINED LAKES
mixed (calcareous), mesic Fluvaquentic Haplaquolls

       deeply fine-loamy, or fine-loamy
       over coarse-loamy (soil LP3)

mixed (non-acid), mesic Fluvaquentic Haplaquolls

       deeply fine-loamy, or fine-loamy over coarse-loamy

mixed (non-acid), mesic Hydric (?) Haplaquolls

       deeply fine-loamy, or fine-loamy over
       coarse-loamy, subsoil non-ripened

mixed (non-acid), mesic Histic Humaquepts

       histic over fine-loamy,
       subsoil non-ripened

mixed, mesic Sulfaquepts

       histic over clayey,
       subsoil non-ripened (soil LP2)

SPECIAL SYMBOLS

       lakes

       boundary between upland and drained lakes

Fig. 16. Simplified soil map of a part of the drained lakes and peat uplands district. Same area as Figure 15; for location see Figure 17.

wood-fen peat carried a forest with alder, willow, ash and some oak, and with some birches further inland. Beyond the area influenced by water-borne sediments were the desolate treeless wildernesses of raised bogs; the places the medieval settlers described by the middle-Dutch word *wildernisse*. The forests of Holland (= holt-land) were, in fact, only gallery forests fringing the small rivers and fen streams in what was otherwise a region of open country.

In the early Iron Age there was probably some sparse settlement on the levees of the Old Rhine (Fig. 17) and alongside the small fen streams. This settlements increased considerably in Roman times. The Old Rhine formed the northern *limes* of the Roman Empire, some placenames on the *Tabula Peutingeriana* can be identified from modern names in this area (e.g. Albanianae = Alphen).

However, the fens remained uninhabited until the tenth century when the Lords of these lands (in the lands to the west of Figure 17 the Count of Holland; in the east, the Bishop of Utrecht) started to grant permission to reclaim (Lambert[1], 1971, p.

1. The original literature in Dutch could be referred to, but Miss Lambert's 'The making of the Dutch landscape' will be more easily accessible to English readers.

102-104). This was done very systematically and regularly: starting from the fen streams, blocks were staked out with fields mostly 6 furlongs long (the Dutch obsolete measure *voorling* has the same meaning (furrow long) and same length (a little over 200 m) and 30 rods wide (there existed several local Dutch *roedes*; this rod is nearly 4 m long). In the course of the 11th and 12th century the fens north of the Old Rhine were reclaimed. Those south of the river were reclaimed in the 13th century. Originally the land was partly arable, this is revealed in old documents: grain tithes had to be paid to the count or the bishop.

The field ditches drained by gravity into the fen streams, but as the soil subsided (due to shrinkage and oxidation) the difference in height between the raised bog and these streams lessened, impeding the drainage, which deteriorated further with the rising of sea level (Lambert, 1971, p. 106). To meet this challenge several Drainage Boards were formed. Over wider regions these were organized into larger units, the High Councils of Landholders (*Hoogheemraadschappen*) the oldest of which existed as early as 1220.

Compared with the clayey wood peat alongside the rivulets, the Sphagnum peat of the former raised bogs with its low ash content was excellent for fuel. Therefore, after the forests had been cut down, peat became the main source of fuel for the cities of Holland. Between 1600 and 1750 a large area was cut over, converting good agricultural land into unproductive swamps and lakes which continuously enlarged as their peat margins were abraded. However, not all lakes in the western part of the Netherlands originated in this way; some have natural causes (Fig. 15). Fen streams were widened into lakes during the marine transgression in the 12th century, particularly in Noordholland province. The man-made lakes are usually called *plas* (Dutch); natural lakes are called *meer* (Dutch) = mere (Eng.).

In these lakes a thin layer of lacustrine mud accumulated over the old marine deposits. Most of this sediment resulted from the abrasion of the lake margins and came from the sod of the original surface which, being unsuitable for fuel, was thrown back by the peat cutters. This mud contained all kinds of remnants of the flora and fauna found in the original lake; after the drainage and reclamation of the

Table 5. Polders reclaimed from shallow lakes in Noordholland, Zuid-Holland and Utrecht Provinces (excluding polders in the former Zuyder Zee)

	Enclosure between									
	1500 – 1550	1550 – 1600	1600 – 1650	1650 – 1700	1700 – 1750	1750 – 1800	1800 – 1850	1850 – 1900	1900 – 1950	1950 – present
Acreage (ha)	82	2228	26 985	1495	3456	21 882	9585	42 007	1648	–
Number of polders	2	17	48	4	7	24	1	68	8	–
Average acreage[1] per polder (ha)	–	147	688	374	494	912	682	735[2]	206	–
Largest polder (ha)	65	620	7 100	960	1125	3 975	4355	18 100[3]	50	–
Smallest polder (ha)	17	5	6	170	270	85	5	4	34	–

1. Excluding polders less than 25 ha.
2. Average acreage excluding the polder of 18 100 ha: 425 ha.
3. Second largest polder: 3015 ha.

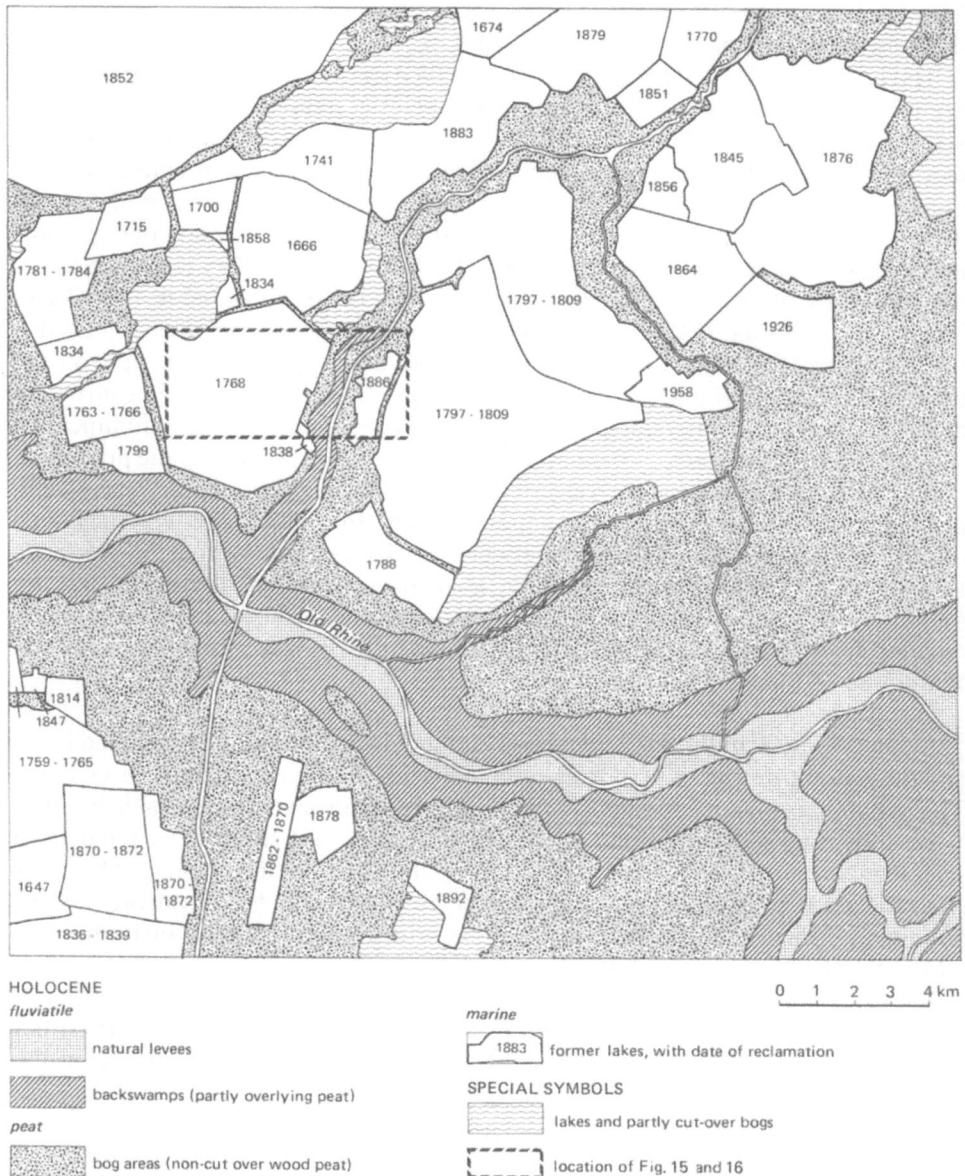

HOLOCENE

fluviatile

natural levees

backswamps (partly overlying peat)

peat

bog areas (non-cut over wood peat)

0   1   2   3   4 km

marine

1883   former lakes, with date of reclamation

SPECIAL SYMBOLS

lakes and partly cut-over bogs

location of Fig. 15 and 16

Fig. 17. A typical part of the drained lakes and peat uplands district. Originally the lakes were raised bogs, cut over for fuel; after drainage an old marine sediment was exposed (3-6 m below sea level). Thus these lakes are man-made geological windows that reveal the surface of four to five thousand years ago. In the northwest is the largest (18 000 ha, 1852) reclaimed lake. Further north, outside the figure, is the Schiphol international airport, situated 4 m below sea level. These former lakes are called bottomlands, and the adjacent undisturbed areas uplands. For location see Figure 3.

lakes, this mud was mixed with the upper part of the marine sediment, resulting in the dark topsoil of these soils (cf. soil LP3, p. 111).

The polders reclaimed from these lakes (called pools by Edelman, 1950, p. 155) are all situated below the 2.5 m below sea level contour (Fig. 3). Reclamation started in the first half of the 16th century (Table 5) with the drainage of some small meres in Noordholland province. It is difficult to be precise about where or when reclamation began, as historical sources differ. The introduction of the wind-driven water mills enabled the lakes to be drained. Although mention of corn mills in Holland is first found at the end of the 13th century (they were probably introduced by returning Crusaders), the water mill is first recorded in 1408 (Lambert, 1971, p. 197 and 124). The efficiency of the wind mills was greatly increased at the turn of the 16th century as a result of two innovations: pivoting the upper part of the mill (the cap on which the sails are mounted) so that the caps could be turned to make best use of the prevailing wind; and using the principle of the Archimedean screw to help lift the water. These improvements greatly speeded up the process of reclamation. Other factors, such as prices of land and agricultural products, wealth of investors, and wars, were also important. The first large lake was reclaimed during the Twelve Years' Truce between Spain and the United Netherlands Provinces (the 7100 ha Beemster Polder, north of Amsterdam, drained in 1608, by 42 mills). The lull in reclamation between 1650 and 1750 (Table 5) may have resulted from events in that turbulent time when there were three Anglo-Dutch Wars, one of them (1672) with France.

A conclusion from this geological and human history is that the reclaimed lakes and the present-day lakes practically coincide with the area of original raised bogs. The remaining uplands were not cut over because wood peat is not suitable for fuel.

To illustrate soil conditions in this district (Fig. 16) three soils have been selected; two from the reclamed lake bottoms and one from the uplands. The latter site (soil LP1, p. 102-105) was sampled north of the Old Rhine on the eastern boundary of Figure 17. The hamlet, on a road with six-furlong-fields on both sides, was mentioned as early as 1308.

The other soils are from reclaimed lakes, soil LP2 (p. 106-109) from the floor of the polder reclaimed between 1836 and 1839 in the southwest of Figure 17; the other (soil LP3, p. 110-113) is from the arable fields south of the lake visible in the northwest corner of the aerial photograph, Figure 15.

Cut-over raised bogs district

On the Soil Map of the Netherlands, scale 1:200 000 (Stichting voor Bodemkartering, 1961) about 450 000 ha have been mapped as peat soils. Also indicated as peat soils on this map are: shallow peat overlying Pleistocene sand; peat overlain by thin marine (Fig. 8) or fluviatile sediments; and peat with an anthropogenic sandy topsoil.

The largest areas of peats lie east of the inland limit of the Holocene marine sediments in western and northwestern Netherlands (Fig. 2). These low-lying peats, which are between 0.5 and 1.5 m below sea level, are traditionally called *laagveen* (low moor). However, these low moors are partly drowned raised bogs, which have been cut over in the west of the country; that area is discussed in the foregoing section. The other peat areas in the Netherlands are mainly raised bogs, traditionally called *hoogveen* (high moor). In my opinion this subdivision into high moor and low moor is obsolete (see discussion on p. 5).

The raised bogs are found in the southeast and the northeast of the country and overlie Pleistocene sands. Nearly all of them are at least partly cut over and the remnants covered with 10-20 cm of sand (p. 118). The two small areas in the southeast are at about 30 m above sea level; in the northeast (Fig. 21) the surface of the cut-over are slopes north from some 15 m above sea level to about 1 m above sea level.

The peat began to accumulate in depressions. The depression in the eastern part of the peat areas of Figure 21 originated as an 'ice-marginal valley' (Ter Wee, 1962, p. 71). This is the *Urstromtal* (German) or *pradolina* (Polish) which ran parallel to the front margin of the Saale continental ice sheet during the last stage that reached the Netherlands, draining the meltwater in a northwestern direction. This valley was partly infilled with fluvioglacial material, with marine deposits in the subsequent interglacial period (Eemian), and with cover sands and a local veneer of loess in the Weichsel ice age. Paludification started locally in late-Weichsel times and peat accumulation only came to a standstill as a result of the drainage preceding and accompanying the peat cutting. The latter 'caused the disappearance of the larger part of the peat, even before there was any question of scientific peat investigation' (Casparie, 1972, p. 5). In the few remnants remaining today, it is not even possible to study the last stage of the bog development 'as a result of the disappearance of a peat deposit, about 1 m thick, due to the buckwheat burning in the 19th and 20th centuries' (Casparie, 1972, p. 251).

The lowest part of the bog floor always has non-podzolized soils, for example the buried mineral soil below soil RB2, p. 121. The first peat formed on these sites is a fen-wood peat, a topogenous eutrophic-to-mesotrophic peat (for description see the DG2 horizon of soil RB2).

As the peat accumulated the influence of the base-rich ground water gradually decreased, the vegetation became solely dependent on rainwater and the fen changed into an ombrogenous, oligotrophic raised bog (for description see soil RB1, p. 117). The peat also spread laterally over the surrounding mineral soils on higher land: on these sites Sphagnum peat overlies a hydromorphic podzol soil (soil RB3, p. 125).

At the end of the sixteenth century only the outermost margins of the area of raised bogs on the Dutch-German border (Fig. 21) had been exploited for peat. The neighbouring villages on the Pleistocene sands with arable land on the plaggen soils, rough grazing land on the common heathlands, and grassland in the small brook valleys (see next two sections), used the adjacent bogs also as common lands (*compascuum*, see Figure 18). This was only possible on the bog margins, and the land use was restricted to a kind of slash-and-burn cultivation system for growing buckwheat and some pasture; furthermore the upper loose Sphagnum moss peat was used as bedding material in the stables (p. 115) and some peat was used for fuel, but only for local consumption.

In the forest-poor Netherlands the ever-increasing demand for fuel, both for domestic and industrial use, initiated peat-cutting on a commercial scale, from the beginning of the sixteenth century. This began, in the north, financed and organized by the citizens of Groningen one of the biggest cities in our country in the 16th century, with 19 400 inhabitants in 1564; later in the middle and in the south,

fig. 18. Part of the cut-over raised bogs district with the villages Emmer-Erfscheidenveen (a) and Emmer-Compascuum (b). These names indicate that their mother-village was Emmen, situated a few km to the west on cover sand overlying glacial till. Before the peat was dug out at the beginning of this century, the area was common land for the farmers in Emmen (L. compascuum = common pasture) which had to be divided :Dutch scheiden = to separate) between the commoners (Dutch erf = heir) of Emmen. Same area as Figure 19.

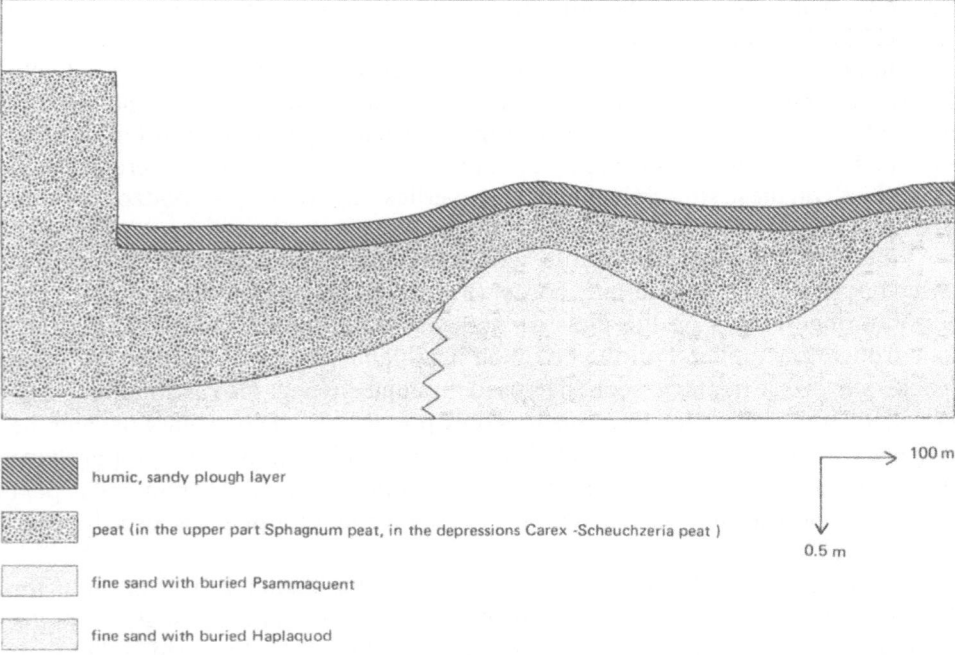

humic, sandy plough layer

peat (in the upper part Sphagnum peat, in the depressions Carex -Scheuchzeria peat)

fine sand with buried Psammaquent

fine sand with buried Haplaquod

100 m

0.5 m

Fig. 20. Sketch-section showing variations in depth of the bog floor and the sub-peat mineral soils. For location see Figure 19.

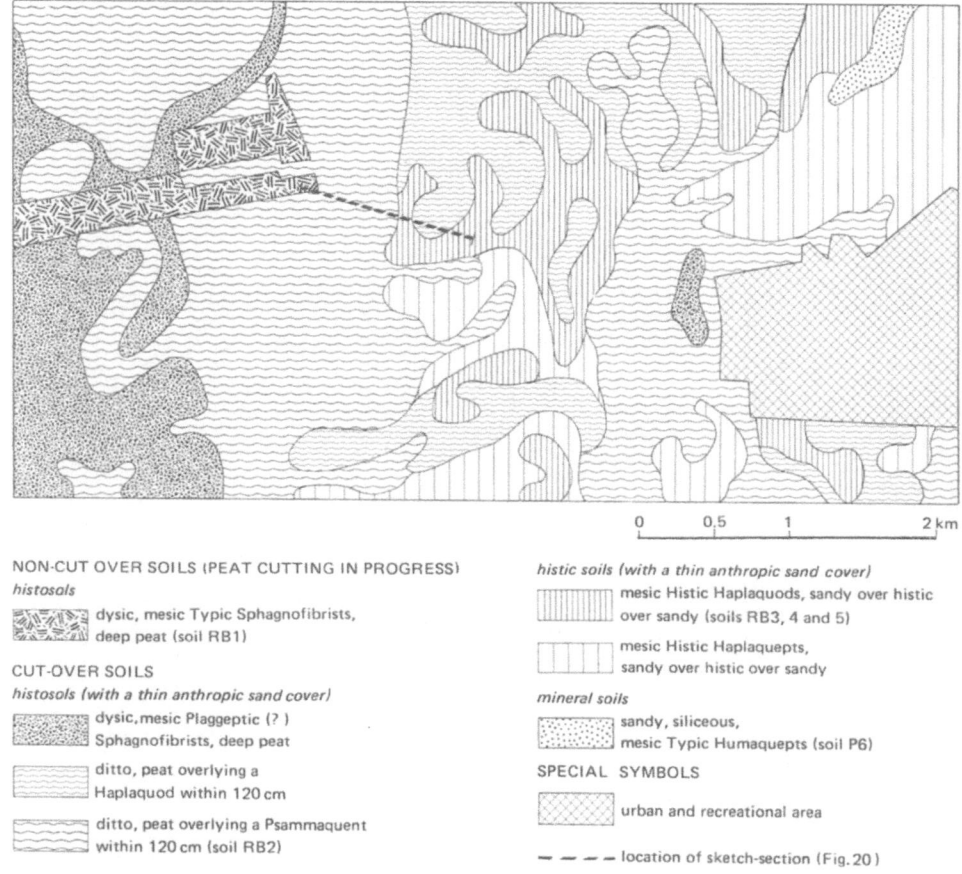

NON-CUT OVER SOILS (PEAT CUTTING IN PROGRESS)

histosols

dysic, mesic Typic Sphagnofibrists, deep peat (soil RB1)

CUT-OVER SOILS

histosols (with a thin anthropic sand cover)

dysic, mesic Plaggeptic (?) Sphagnofibrists, deep peat

ditto, peat overlying a Haplaquod within 120 cm

ditto, peat overlying a Psammaquent within 120 cm (soil RB2)

histic soils (with a thin anthropic sand cover)

mesic Histic Haplaquods, sandy over histic over sandy (soils RB3, 4 and 5)

mesic Histic Haplaquepts, sandy over histic over sandy

mineral soils

sandy, siliceous, mesic Typic Humaquepts (soil P6)

SPECIAL SYMBOLS

urban and recreational area

— — — — location of sketch-section (Fig. 20)

Fig. 19. Simplified soil map of a part of the cut-over raised bogs district. Same area as Figure 18; for location see Figure 21.

financed by capital from the western provinces and from merchants in Amsterdam. Within 350 years almost all the area of raised bogs delineated on Figure 21 had been cut over; the only small remnant left today is in the south, and it is even smaller than the area depicted on the map, which is based on data from a survey made between 1953 and 1955.

In order to be cut, the peat must be drained. To transport the *turf* (this Dutch word is only used for the dried peat that is intended for fuel) by barge, canals are required. Consequently the area shown in Figure 21 is characterized by a network of waterways, consisting of one rather large canal and many smaller laterals and ditches. The large canal runs from southeast to northwest in the bog and has many locks. Its name, the *Stadskanaal,* refers to the *stad* (town) of Groningen, which ordered the canal to be dug in order to open up the vast peat bogs southwest of the town. The *Stadskanaal* has many side-canals, called *hoofddiepen* (head-deeps); in some cases two of them run closely parallel to each other thus creating a strip of land where the labourers from the peat-cutting companies originally lived and where the villagers later settled (Fig. 18). Narrower laterals, called *wijken,* were dug perpendicular to the *hoofddiepen.* They were spaced 100-250 m apart, with a ditch halfway in between. The distance between the *wijken* shown on the aerial

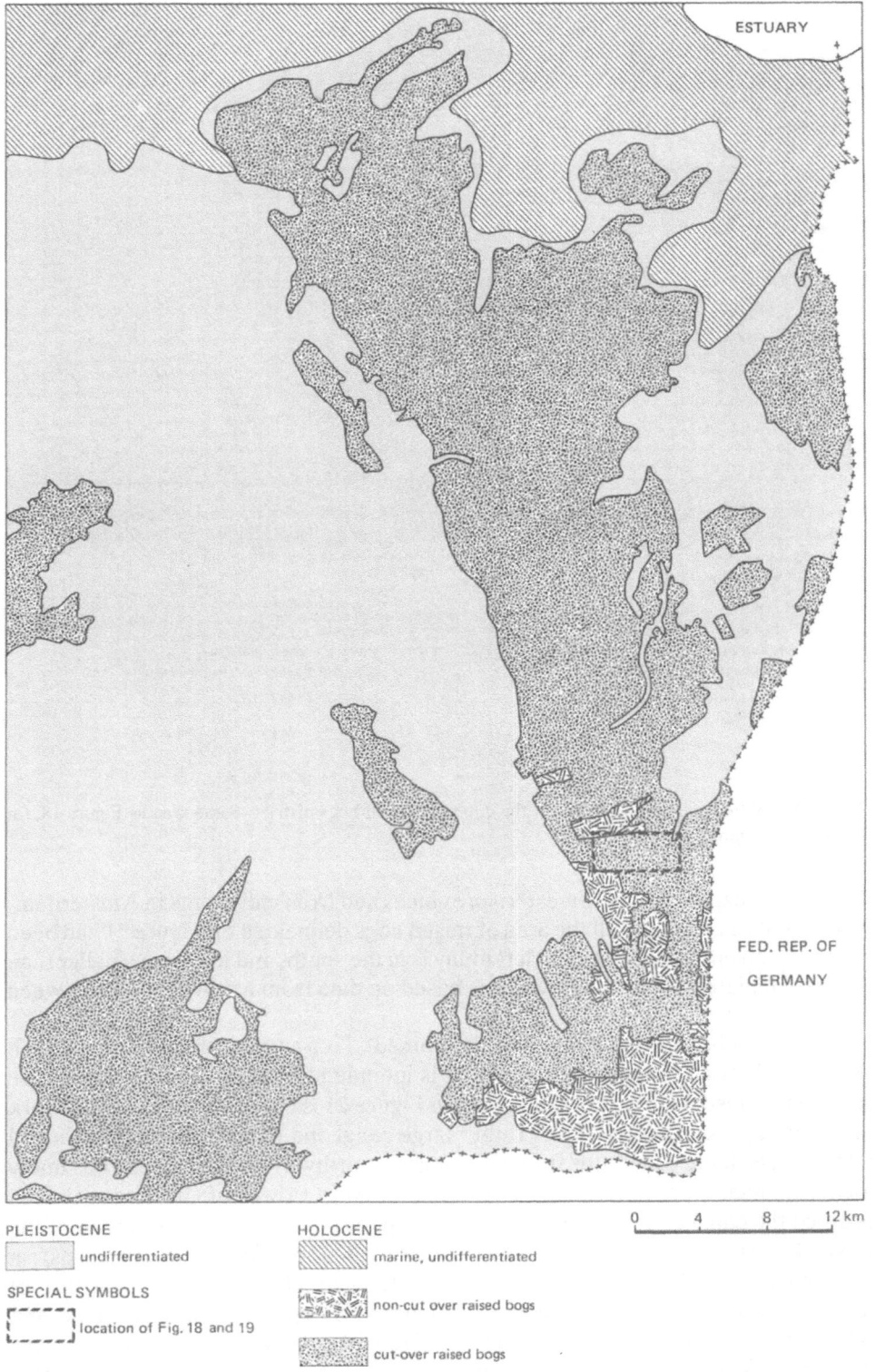

Fig. 21. Main area of the former raised bogs in the Netherlands. For location see Figure 3.

photograph is 200 m, so the individual fields are 100 m wide (cf. the section 'Drainage and ground water' on p. 118).

Not all the peat was removed; the lower peat was often not excavated because of drainage difficulties, and the upper loose peat (the C2 horizons of soil RB1, p. 117) being unsuitable for fuel was thrown back into the turbary (the Dp horizon of soil RB2, p. 121). These peat remnants were levelled and dressed with spoil from the adjacent canals, which were dug into the sandy bog floor.

The resulting soils were poor, being acid as well as low in nutrients, but were physically good: a well-regulated water level and the spongy moss peat acted as a wick and drew up the water, in fact a kind of hydroponics. The first reclamations relied on the night-soil and street refuse of the town of Groningen; later well-balanced fertilizer dressings, including lime and micro-nutrients were used.

The farmers settled along the main and secondary canals, and therefore the settlements are linear. The reclaimed areas (comprising land and settlement) are called *veenkoloniën* (peat colonies). The new villages ware given different types of names. Some were given fanciful names (Erica, founded 1912); some were called after people who played an important role during the first days of settlement (Wildervank, founded 1647); others after the neighbouring mother-villages (Fig. 18); and, as in other settlements in the colonies overseas, some were named after cities in the old land and given the prefix 'New'. In 1851, in an area where the peat-cutting had been financed by Amsterdam money-lenders, a village Nieuw-Amsterdam was founded (population 4416 in 1970) some two hundred years after the foundation of an earlier colonial New Amsterdam in the New World!

Figure 19 illustrates the soil geography of a typical *veenkolonie;* the sketch-section (Fig. 20) shows that the differences in the soils are mainly controlled by the depth and the type of the bog floor.

Soil RB1 (p. 117) is an example of a peat that has not been cut over (though a part of the loose moss peat may have been removed for garden use); soil RB2 (p. 121) has been partly excavated (a thickness of several meters of peat may have been used for fuel). The other three soils (RB3 on p. 125, RB4 on p. 129 and RB5, on p. 133) are examples of so-called *versleten veenkoloniale gronden* (wasted peat-colonial soils, cf. p. 127); the latter two have been mixed.

Anthropogenic soils

This is the only section in which the soils are not arranged according to their geographic relationships and surface geology. The five anthropogenic soils (soils A1-A5, p. 134-153) are grouped together because they share one dominant soil-forming factor: man. Therefore it is impossible to correlate such soils with a particular parent material on the map (Fig. 2).

Depending on the definition used, anthropogenic soils can be found in any of the soil regions of the Netherlands because all the soils are influenced by man, to a greater a lesser degree. Even our few 'virgin' soils (virgin in the sense of never having been ploughed or fertilized) are affected by man. In the Humod (soil P3, p. 165) 'the black B2h is due to heath vegetation' (Edelman, 1950, p. 59) and the heath vegetation is anthropogenic: the original forest was over-exploited and therefore was replaced by heath (see below). Another example is the Hydraquent (soil M1, p. 65) in which the sedimentation is not entirely natural: brushwood groynes at the

Fig. 22. Plaggepts or Plaggen soils (a) situated between inland dunes with Udipsamments (b) ih the east and hydromorphic soils with Humaquepts (c) in the west. The Plaggepts are arable, the inland dunes are partly stabilized by forest, partly still active dunes, the Humaquepts are all grassland. The recreational value of the dune land is illustrated by the presence of some camp sites (d). Same area as Figure 23.

junction of the tidal silt and the tidal sand flat impede the tidal currents and promote settling of the sediment (Kamps, 1963). Embankment, drainage and subsequent oxidation changes the soft blue mud of the forelands into the firm grey soils of the polders (such as soil M6, p. 85).

Liming, fertilizing, ploughing, and so on, are considered to be 'normal' anthropogenic modifying factors in the Netherlands; deep ploughing and subsoiling modify the soil to a greater degree. For such soils, which 'have been deeply mixed by plowing, spading or other moving by man' (SSS, 1975, p. 187) there is a specific class in Soil Taxonomy, called Arents; soils RB4 (p. 129) and RB5 (p. 133) are examples of such Entisols, which are discussed with their undisturbed prototype in the preceding section.

However, there is a tendency in the Dutch Soil Survey Institute to consider soils as being 'man-made' only if they have beeñ gradually raised by the addition of mineral material by some agricultural practice. Such upper horizons are coded Aan (SSS, 1951, p. 182), a letter subscript which, unfortunately, has been dropped in Soil Taxonomy. The original soil is considered to be buried when the Aan horizon is thicker than 50 cm; such a soil is then classified as a 'thick earth soil' in the Dutch system (De Bakker & Schelling, 1966, p. 182). Examples of soils with thinner man-made topsoils are soil LP1 (p. 105) and soil P2 (p. 161).

In this section two plaggen soils (soils A1 and A2, p. 137 and p. 141) are discussed, together with the inland dune (the Udipsamment, soil A3, p. 145). The three can be found in the Pleistocene sandy district, and although the Udipsamment is no anthropogenic soil in the above-mentioned sense, it is discussed in this section together with the plaggen soils because it is related to the plaggen soils historically as well as geographically.

The other two anthropogenic soils occur in the marine district. Both have also

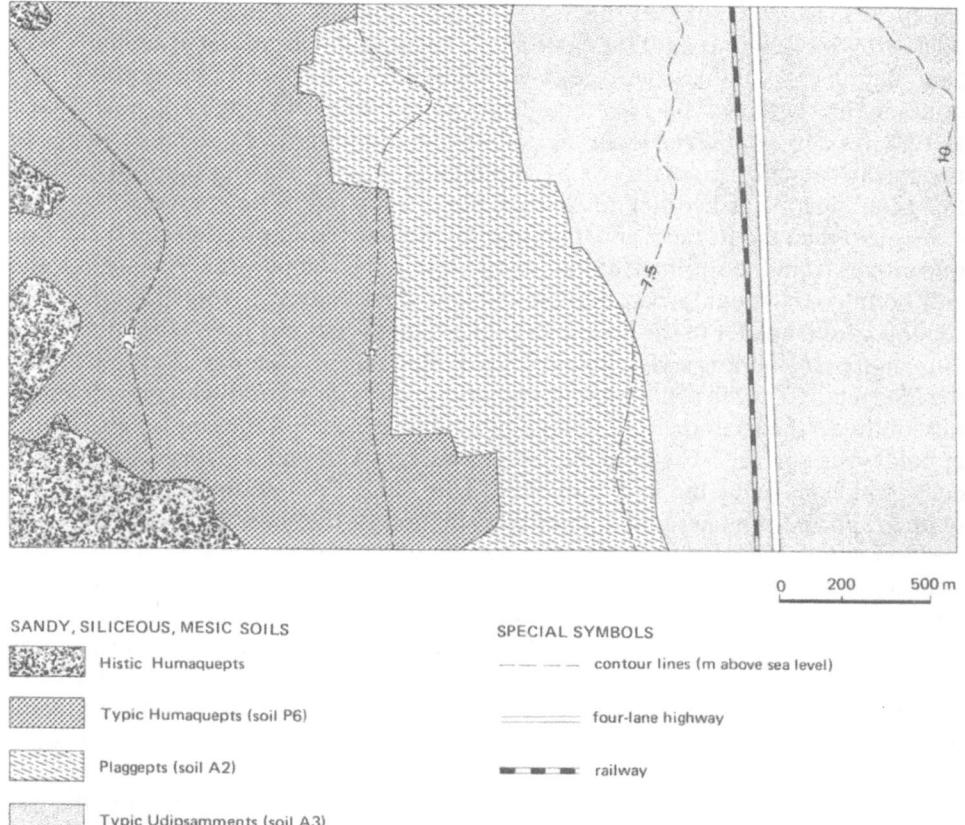

0   200   500 m

SANDY, SILICEOUS, MESIC SOILS

Histic Humaquepts

Typic Humaquepts (soil P6)

Plaggepts (soil A2)

Typic Udipsamments (soil A3)

SPECIAL SYMBOLS

———— contour lines (m above sea level)

======== four-lane highway

▬ ▬ ▬ railway

Fig. 23. Simplified soil map of a part of the Pleistocene district with two important anthropogenic soils, one man-made (the Plaggept) and one man-induced (the Psamment). Same area as Figure 22; for location see Figure 24.

been raised gradually by man; one (soil A4, p. 149) by using dune sand and mud dredged from the ditches, and the other (soil A5, p. 153) by taking material from the margins of the individual fields and bringing it towards the centre, thus creating crested or domed fields.

Plaggen soils and inland dunes. These soils are found in the Pleistocene sandy district and they are therefore also discussed in next section. Some general information on their soil pattern, history of land use, agriculture, and so on, can also be found in that section.

In this section more details will be given about the genesis of plaggen soils and the reasons why these soils are often associated with inland dunes.

In most places in the Pleistocene sandy district this relationship between plaggen soils and inland dunes is obscured by the complicated soil pattern which is mainly controlled by the relief (e.g. Fig. 28 and 31). However, in the sample area (the aerial photograph with soil map, Figures 22 and 23) this relationship is quite clear. The area shown on the photograph is situated at the foot of the western slope of the hills in central Netherlands (Fig. 3). These hills consist of coarse fluviatile preglacial sands, pushed up into hills by the Saale ice sheet; the lower slopes are covered with coarse fluvioglacial sands of the same age. The upper 2-4 m consist of aeolian cover sand dating from the last ice age (Weichsel), which has been blown into inland dunes in the southeast[1] part in recent times (Steur & De Bakker, 1969, p. II-33 and II-125). As can be inferred from the contour lines on Figure 23, the area slopes to the northwest with a gradient of 2.5-3 m/km, and there is no superimposed relief on this general slope as in other cover-sand areas (cf. p. 48). The ground-water level slopes less than the surface, and therefore the watertable approaches the surface as one moves from the southeast to the northwest in the sample area. As a result, the soil boundaries roughly parallel the contour lines: moving from the area with shallow ground water to the area with deep ground water one meets first the Histic Humaquepts (Half Bog soils), then the Typic Humaquepts (Humic-Glei soils), then the Plaggepts (plaggen soils) which have an intermediate ground-water level, and in the southeast the Psamments (inland dunes) where the ground water is more than 3 m below the surface. The boundary between the 'wet' and the 'dry' soils practically coincides with the old main road (Fig. 22). Further to the right on the photograph the modern traffic lines are situated in the dune area, avoiding the built-up areas; the left one is the railway, the other two are the E35, a motorway with separated double lanes. These three roads link western and northern Netherlands. Most of the farmsteads are situated along the old road, which was originally of local importance only. This location enabled the settlers to have their grasslands on the relatively nitrogen-rich Humaquepts (such as soil P6, p. 177, cf. also p. 49), and their arable land on the soils with an intermediate hydrology; uphill, beyond the arable land, lay the common heathlands on non-hydromorphic podzol soils, such as soil P3, p. 165.

Some of the consequences of this hydrological situation are: the fields in the left of the photograph have water-filled ditches; near the stately house in the middle of the photograph the shallow watertable has been exploited to create ornamental lakes; the arable land has baulks as field boundaries and neither it nor the inland dunes have open drains. In addition, the location of the farmsteads permitted the original settlers to have shallow wells, an essential condition for settlement before the introduction of piped water.

In most regions of the Pleistocene sandy district, agriculture in the last century was subsistence farming, being more or less in equilibrium with its environment. Only a few products left the farm (some rye and wool, but mainly eggs and butter to be sold in the local market-towns). The heathlands, however poor, constituted an essential element in the system. They were the source of the heather sods that were used as bedding material in the cowsheds and sheepfolds (Pape, 1970). An article by Domhof (1953), discussing and illustrating how the sods were cut, contains a photograph of the special scythe and pitchfork used to peel the thin heather sods. These turves were also used to cover the margins of active dune land in an attempt to check the advance of the shifting sands. Both these uses are frequently mentioned in nineteenth-century Reports of the Ministry of Agriculture; another use for heather, namely broom-making is referred to in these Reports. In 1898 132 000 brooms were made in the village of Oldebroek (in the northeast of figure 24) and were peddled at 2 cts each in neighbouring market towns and even 'exported' to the

1. Contrary to all other Figures, which are oriented south-north, the Figures 22 and 23 are oriented southwest-northeast (cf. Fig. 24).

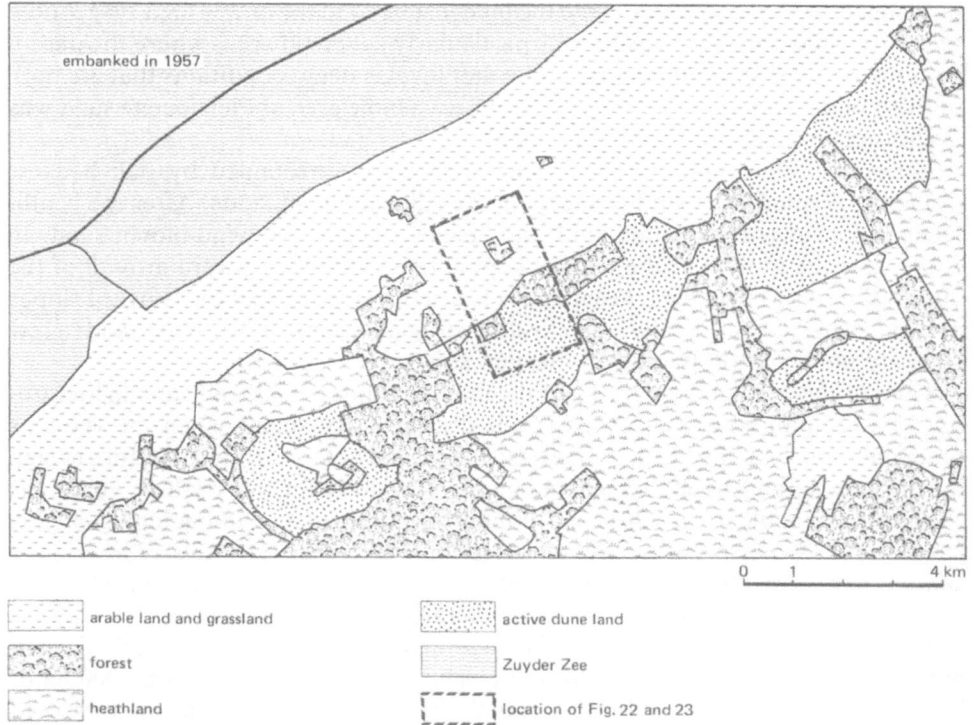

embanked in 1957

0 1 4 km

arable land and grassland

forest

heathland

active dune land

Zuyder Zee

location of Fig. 22 and 23

Fig. 24. Land use in 1870. For location see Figure 3.

more urbanized western part of the Netherlands. This income represented indeed a substantial amount of money to the moneypoor community of those times.

Perhaps the most important use of the heathlands was its use as rough grazing land for a special breed of sheep adapted to local conditions. The flock had to spend every night in the sheepfold so that all the droppings could be collected. In these pens, and also in the cowsheds, which had a special kind of stable (called pot-stable by Edelman, 1950, p. 24) heather sods were used as bedding material. These sods consisted of the upper organic-rich layer of the podzol soils with some adhering sand. The dung-impregnated bedding was used to manure the arable land; the mineral part of this earth-containing manure stayed back as a kind of weathering residue. Consequently the arable fields were raised very gradually (cf. p. 134), changing the podzol soil into a plaggen soil.

Based on some assumptions (amount of plaggen needed to make manure for one hectare of arable land; area of heathland where these plaggen had to be cut; time of rest to restore the organic layer) Slicher van Bath (1963, p. 258) calculated that this alone needed 3-7 ha heathland. To provide this quantity of manure two cows or 20-30 sheep were needed: taking into account the grazing area required for this livestock 10 ha of grassland and heathland were needed to manure one hectare of arable land (Edelman, 1950, p. 23).

The over-use of heathland (de-sodding, heath-mowing for brooms, sheep-grazing and fires) and the daily trek of the sheep flocks to and fro from the farmsteads to the heathlands made the land just beyond the arable area susceptible to severe wind erosion. Erosion was particularly severe in areas where the parent material is cover sand and the ground-water level is deep; conditions that are both fulfilled in the area shown in Figure 24, where a large area of Pleistocene sand was again transformed into active dune land.

Formerly, all complexes of plaggen soils were surrounded by oak hedges, originally planted to retain the sheep on the adjacent heathlands. After the heathlands had changed into dune land, the hedges trapped the wind-blown sand. In some sites the growth of the hedges kept abreast with the upward growth of the dunes; the slope of such enclosed dunes is usually steeper than the leeward slopes of wandering dunes. Locally, man lost his struggle against the moving sand; a few sites are known with overblown plaggen soils.

Tales of the ever-present threat of the moving sands exist locally and in literature: for example, how poor crops resulted after a dry spring when the rye was practically shorn off by bleak strong east winds, and how a son had therefore to sign for the crew of an East Indiaman and had to stay away for two or three years, his remittances enabling the farm and familiy to survive. At the beginning of this century the Swedish author Selma Lagerlöf wrote a book (*Niels Holgerssons underbara resa* = The wonderful adventures of N.H.); in which one of the stories is about a farm in southwestern Sweden, threatened by shifting sands.

The sites bordering dune land are characterized by humus-poor, thick plaggen horizons, because of dilution with dune sand. Most of the original 80 000 ha of inland-dune land is now stabilized, mostly by Scots pine. Only a few thousand hectares of active dune land remain today and they are scattered in small areas, few of which exceed 100 ha. Today the situation is reversed, for we are now trying to keep the sands moving in order to preserve some active dune land.

The plaggen soils found in the landscape described above have black (dark grey) plaggen horizons containing many bleached sand grains, which have a rather high C-N ratio. These features are derived from the humus podzols of the heathlands that supplied the plaggen (Edelman, 1950, p. 24; Domhof, 1953, p. 203; Pape, 1970, p. 252; Mückenhausen, 1975, p. 453 and 454; Mückenhausen, 1977, p. 137 and 138). The brown plaggen soils are situated on cover sand ridges in or bordering small river valleys (cf. Figure 13 in: Pape, 1970), their topsoil is more clayey and their C-N ratios are lower than those of their black counterparts (cf. the analytical data on p. 136 and 140). Most authors agree that 'the use of clayey sods, taken from this valley, has caused the brown colour' (Pape, 1970, p. 252), but some authors are more cautious and say this 'is not quite so evident' (Domhof, 1953, p. 203) or: 'it is generally assumed' (De Bakker & Schelling, 1966, p. 191).

As far as we know, plaggen soils are restricted to only part of the glacial plains of northwestern Europe (Fig. 25), and only occur in areas with fluvioglacial sands or cover sands. They are not known to occur in the sandy western part of Denmark and Schleswig-Holstein, or, apart from a few insignificant areas (see p. 135) in the sandy northern part of the German Democratic Republic, and Poland, although all these areas have analogous soil conditions. I would hazard a guess that this agricultural practice was restricted to the Germanic tribe of the Saxons and partly of the Franks, and that the Frisians, the Scandinavians and the Slavs did not use this practice.

Fig. 25. The distribution of Plaggen soils in northwestern Europe.

Raised soils from marine parent material. These two soils are situated in what has been called 'the older land' in the first section of this chapter; one comes from the southwest (south of The Hague, soil A4), and the other from the north (about 30 km north of the northern boundary of Figure 8; soil A5). Although these soils are not plaggen soils in the classic sense, they fit the new concept of the Plaggepts in Soil Taxonomy, because the surface horizon satisfies the criteria for the Plaggen epipedon (SSS, 1975, p. 18 and 257). Likewise the sanded soils in Ireland are not classic plaggen soils, neither are the other man-made soils Conry (1974) mentions in his review article.

Soil A4 (p. 149) is situated in the Westland. This is the most important horticultural district in the Netherlands; it lies in the triangle The Hague-Rotterdam-Hook of Holland. The district is bordered by coastal dunes in the west, and by the *Nieuwe*

Fig. 26. The central part of Westland, the horticultural district between Rotterdam and the Hague (a); soil A4 has been sampled at (b). The white rectangles (c) are large, modern glasshouses mostly used for growing flowers (chrysanthemums, freesias and carnations), white-washed to prevent sunburn on this bright spring day (22 April); in the small glasshouses (d) grapes are grown, in the other glasshouses lettuce, tomatoes and cucumbers are important crops. At (e) one of the three auction buildings for vegetables in the Westland (the three had a turnover in 1975 of Dfl. 456 000 000; the only flower auction in the Westland Dfl. 363 000 000). The small river Gantel (f) is a fossil tidal creek, associated with the marine transgression of the fourth century A.D. which submerged and buried Roman and native settlements (g) that were rediscovered during the soil survey of this area. For location see Figure 3.

Waterweg (the shipping lane from Rotterdam to the North Sea) in the south. The soils are derived from marine parent material; further inland there are peat soils, interspersed with silted-up tidal creeks, which partly covered the peat with a thin marine deposit. The time of empolderment is unknown, but is probably earlier than the eleventh century. A large part of the area of mineral soils is used for horticulture, both in the open and under glass (Fig. 26); in the peat hinterland there are dairy farms.

The climate and location of this district are advantageous. The climate is slightly more Atlantic than further inland: there are fewer late frosts in spring, fewer hailstorms in summer, and more hours of sunshine. The presence of large markets nearby (Rotterdam, The Hague) encouraged the development of horticulture. As early as 1712 parts of this area were indicated as orchards and vegetable gardens on a map; today the flowers and vegetables are destined for more distant markets and the short distance to Rotterdam airport and harbour is advantageous. The growers are organized into so-called auction-societies. All products have to be sold in auction-buildings; in this district there are three for vegetables and one for flowers (with a total turnover of Dfl. 820 000 000 in 1975).

Formerly nearly all transport was by water. It was possible and necessary to reach the auction-buildings by barge, because the produce had to be brought by barge through the auction hall in front of the auctioneer and the wholesale dealers.

The same barges were used to ship calcareous sand from the inner side of the coastal dunes to the horticultural fields. The sand was mixed with manure from the dairy farms in the neighbouring peat district and with mud dredged from the ditches and canals and used as manure. Consequently the fields became raised and the soil texture was changed from a silt loam to a loamy sand (cf. the analytical data on p. 148). These loamy sand soils are 'earlier', i.e. produce a crop earlier in the year, when prices are higher; they are also easier to work, e.g. to ridge into asparagus beds (an important crop at the turn of the century). The excavated dune land that supplied the calcareous sand is now at the same level as the adjoining marine polders, and is also used for horticulture (bulb-growing).

Nowadays the practice of excavating sand for manure has been abandoned because of high labour costs and because practically no area remains where it is allowed to level the dunes by excavation (the dunes act as a sea wall here!). Many of the canals have been infilled and the horticultural produce is transported by road.

Soil A5 (p. 153) is situated on a coastal barrier in the north of the country, described in the section 'Marine district', sub-section 'the older land'. The soils of these coastal barriers have little clay, so they can barely crack; the sand separate is very fine: there are hardly any grains over 0.01 mm in size (cf. the Analytical data on p. 152). Such soils have a low permeability and a low storage capacity, therefore they need to be artificially drained. Today that is no problem: such fields can be tile-drained. But before the introduction of earthenware pipes or of the slotted plastic pipes of today, the farmers had to use an open-drain system, consisting of shallow furrows alternating with slightly elevated beds (such fields today are tile-drained and the old furrows were used for the tiles). The relics of such systems can be seen on many aerial photographs especially when polders are partly sub-merged by inundation (Zonneveld, 1960, p. 44; De Bakker & Marsman, 1978), because there are still slight differences in height between the former beds and the former furrows.

In the northern marine area this pattern is not found, because individual fields were not subdivided into beds with furrows between, but the field as a whole was domed by moving soil material from the margins and the corners to the middle of the fields, thus enabling a type of surface drainage. Such fields are called *kruinig* (Dutch) by the farmers; this translates literally as 'cresty' (De Bakker & Marsman, 1978). Soil A5 (p. 153) was sampled on the crest of such a field. On some fields that were originally superficially decalcified, this practice has resulted in a peculiar situation in which edges and corners of the domed fields have calcareous soils, whereas the soils on the crest are slightly acid (Fig. 38, p. 151).

Such fields are also known northeast of Ghent in Belgium. They were pointed out during one of the Post-Congress excursions of the Fourth International Congress of Soil Science in 1950. In the excursion guide these fields with a convex surface are called 'rounded'. The explanation might be the same as in the Netherlands: surface draining, but Snacken (1971) supposes that liming with calcareous material from the bottom of the surrounding canals is the major cause.

Pleistocene sandy district

Soils derived from Pleistocene sandy materials occupy about 1.3 million hectares in the Netherlands, or roughly 40% of its area. On the map of the surface geology of the country (Fig. 2) three mapping units have been delineated:
1. fine and loamy fine sand,
2. ditto overlying loamy material,
3. sand and coarse sand, often gravelly.

Most of the Pleistocene sands, namely about one million hectares, are fine and loamy fine sand, called cover sand, an aeolian deposit of the Weichsel Ice Age (Table 1, Sample 7a and 7b). Soils P1 (p. 157), P3 (p. 165) and P6 (p. 177) developed from thick cover sands; soil P5 (p. 173) is underlain at 145 cm depth by coarse fluviatile sand.

The second mapping unit occupies a much smaller part of the country; about 140 000 ha. The upper part is also cover sand; the loamy material in the subsoil is mainly glacial till from the Saale Ice Age (Table 1, Sample 8a). Soil P2 (p. 161) was sampled in the north of the country: it developed in cover sand overlying glacial till at 75 cm depth.

The third subdivision in the Pleistocene sandy district consists of thick, sandy, mostly gravelly sediments, which are much coarser than the cover sands. The main area lies in the centre of the country and consists of preglacial fluviatile sediments shaped into low hills by the Saale ice sheet. The Brown Podzolic soil (soil P4, p. 169) came from such a hill west of Arnhem.

Only the first-mentioned part of the Pleistocene sandy district, the area with deep sandy parent material, will be discussed in greater detail.

These Pleistocene sands slope very gradually (0.5-1 m/km) towards the Holocene area (the main river valleys, Figure 31, and the marine polders). The groundwater level has almost the same gradual slope; its fluctuation (the difference between summer and winter level) is often approximately 1 m, and in winter it is on many sites less than 1 m deep (Knibbe, 1969, p. 63).

Superimposed on this general relief and hydrology there is a detailed topography that gives rise to differences in hydrology and controls the soil pattern, the land use, settlement and reclamation history. The cover sands have a typical topography of ridges and depressions of aeolian origin, elevated 1-2 m above and 0.5-1 m below the general surface level, respectively. Shallow valleys (Fig. 31) are numerous: they were most probably eroded by melt-water after the sands had been deposited in the last ice age (Weichsel).

The area is drained naturally by brooks that run in these shallow valleys; locally these are partly infilled with sedge-wood peat, while the depressions are infilled with oligotrophic peat. Today, nearly all the land has been reclaimed and is drained by numerous ditches that discharge into the brooks; the depressions that originally lacked natural surface drainage are now also connected via ditches to the natural waterways in the valleys.

The relative elevation (above the sloping ground-water level, cf. Knibbe, p. 61, 1969) rather than the absolute elevation, defines whether a soil is 'low' or 'high' – the farmers' expressions for hydromorphic and non-hydromorphic soils.

Figure 31 illustrates an area as sketched above; it slopes from the ice-pushed ridge in the east to the floodplain of the river Yssel in de west. The ice-pushed ridge rises about 60 m above the cover-sand area; the hill is partly afforested and partly still heath-covered. The soils are non-hydromorphic podzols, developed in coarse textured, gravelly material, deposited by the pre-Saale rivers, and pushed into low hills by the Saale ice sheet. From the contour lines on Figure 31 it can be deduced that the slope of the cover-sand are is very gradual; the gradient is 1 m in 1200 m. Such a slope is not visible in the field, but its direction can be detected from the water flow in the numerous brooks in the shallow valleys.

The soil map (Fig. 28) shows the soil conditions of a part of the area depicted in Figure 31. The valley soils are mostly Typic Humaquepts (Humic-Glei soils, such as soil P6, p. 177), and the relatively lowest sites have shallow peat overlying sand (Histic Humaquepts of Half Bog soils). On the cover-sand ridges, which locally attain 2-2.5 m above the adjacent valleys, there are Plaggepts (plaggen soils, such as soil A2, p. 141). There are two sites with stabilized dunes (Typic Udipsamments, such as soil A3, p. 145): these two kinds of soil are discussed in detail in the preceding section on Anthropogenic soils. On lower sites there are soils with a man-made surface layer that is too thin (30-50 cm thick) to qualify as a plaggen epipedon (SSS, 1975, p. 18). Such soils are named Plaggeptic (?) Haplaquods (the adjective indicates that they are considered to be intergrades between Plaggepts and Haplaquods; the question mark signifies that as yet there is no such subgroup in Soil Taxonomy). The Plaggeptic (?) Haplaquods occur also as small inclusions in the middle of non-raised soils, such areas are delineated on the soil map (Fig. 28) as a soil association of units 3 and 4. In the sample area non-raised Typic Haplaquods (of which soil P1, p. 157, is a typical example) are the most extensive soils. Although all Typic Haplaquods are hydromorphic (no iron coatings on the sand grains) the soils in the mapped area show differences in drainage class resulting from small variations in elevation and texture (fine sand and loamy fine sand). On the Soil Map of the Netherlands, scale 1:50 000, from which Figure 28 has been taken, these drainage classes are indicated as so-called water-table classes (Van Heesen, 1970); in this Figure these classes have been omitted.

In the discussion about the Humaquept (soil P6, p. 175) it is stated that the relief is the decisive factor in soil formation; it defines the differences between the Humaquepts and the Haplaquods.

Apart for some small pine-woods, the whole area is arable or grass (Fig. 27), in marked contrast to the land use of a century ago (Fig. 29). If the land-use map (Fig. 29) is compared with the soil map (Fig. 28), it can be seen that the area of grassland practically coincides with the Histic and Typic Humaquepts; in addition the area of heathland coincides with Typic Haplaquods. The correspondence between soils and the land use in last century can be explained by differences in the natural fertility of the soils in the pre-fertilizer era, which in turn partly explains the agricultural system of those times.

The natural nitrogen level of the Humaquepts is much higher than of the Haplaquods (Knibbe, 1969, p. 92): the C-N ratio of Humaquepts usually lies between 10 and 16. In the turf of a Haplaquod that has been fertilized for over 50 years (p. 156) the C-N ratio is 12.4, but directly below the turf the ratio is 17.5; C-N ratios higher

Fig. 27. Part of the Pleistocene sandy district east of the river Yssel. Since last century (cf. Figure 29) all heathland has been reclaimed for arable and grassland (the forest (a) was already forest in 1882). The small areas (b) with inland dunes (Udipsamments) are afforested; the Histic Humaquepts are used exclusively as grassland (c); the Typic Humaquepts predominantly as grassland (d); the other soils are used both as grassland and arable land. Same area as Figure 28 and 29.

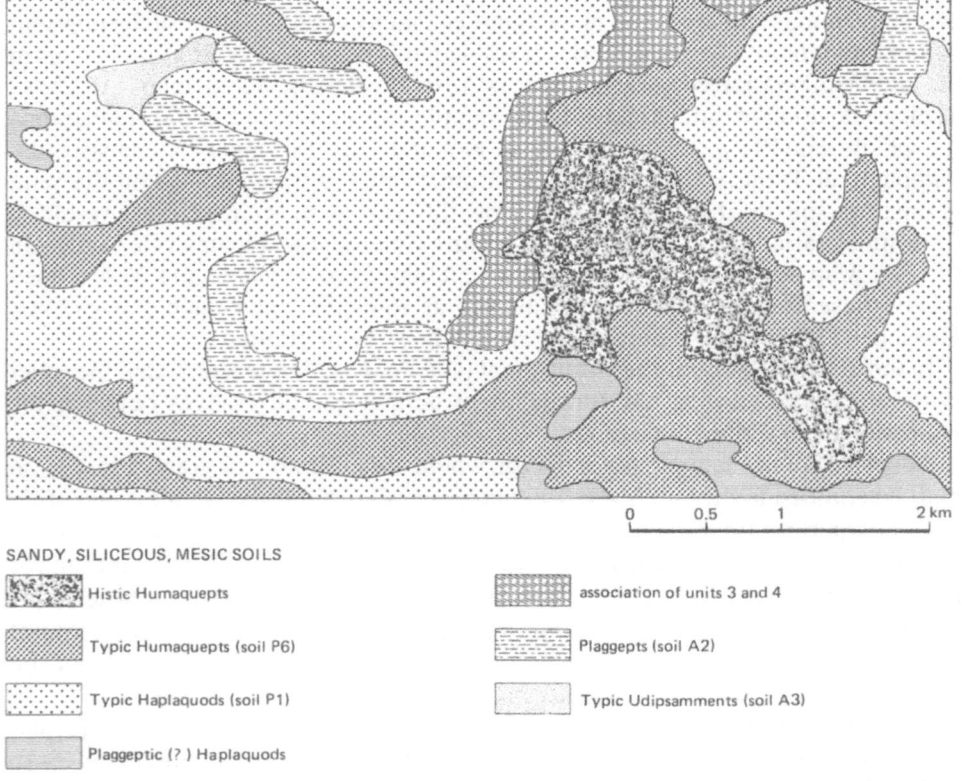

0   0.5   1   2 km

SANDY, SILICEOUS, MESIC SOILS

Histic Humaquepts   association of units 3 and 4

Typic Humaquepts (soil P6)   Plaggepts (soil A2)

Typic Haplaquods (soil P1)   Typic Udipsamments (soil A3)

Plaggeptic (?) Haplaquods

Fig. 28. Simplified soil map of a part of the Pleistocene sandy district. Same area as Figure 27 and 29; for location see Figure 31.

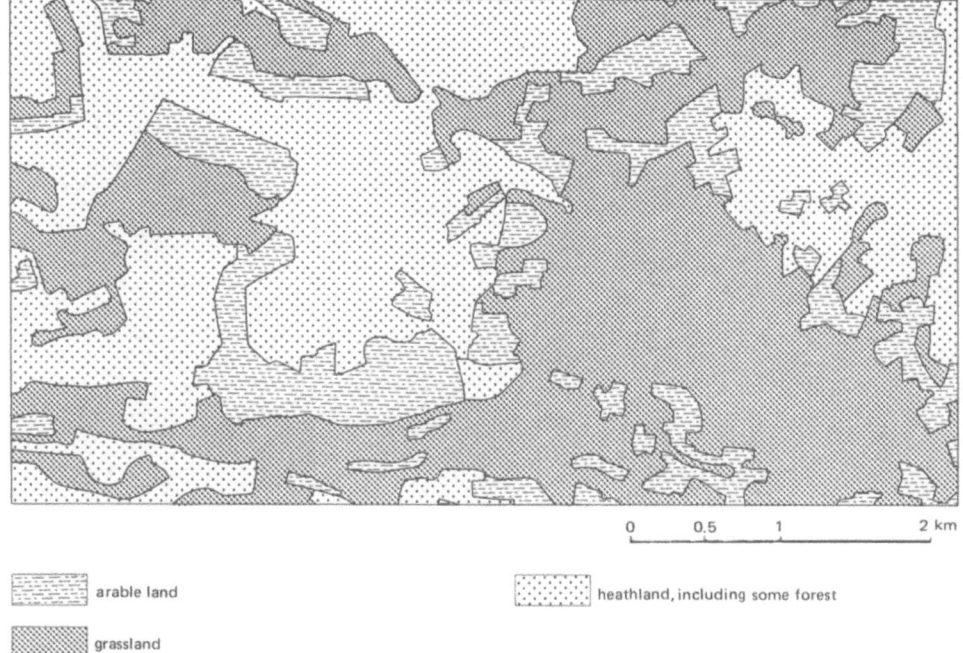

0   0.5   1   2 km

arable land   heathland, including some forest

grassland

Fig. 29. Pattern of land use in 1882 of the same area as Figure 27 and 28. There is a striking resemblance between the land use of that time and the soil conditions (cf. Figure 28): heathland coincides with practically all the Typic Haplaquods and the Typic Udipsamments; grassland with the Typic and Histic Humaquepts; arable land with the other soils.

than 30 have been found in virgin Haplaquods. Most probably this difference in nitrogen level results from the differences in the primeval vegetation (cf. p. 175): an alder forest on the Humaquepts and a birch forest on the Haplaquods. Alders have root nodules with Rhizobia and thus can fix nitrogen. The Haplaquods were (and most of them still are) more acid than the Humaquepts (cf. p. 156 and 176); even today the pH figures differ by a full point. These differences are all probably determined by relief: an ombrogenous situation for the Haplaquods and a topogenous situation for the Humaquepts (see also the discussion on the Humaquept: soil P6, p. 175).

The common heathland was used as rough grazing land for sheep; the grassland in the valleys on the Humaquepts was used as pasture for cattle. The heather also supplied the *plaggen* (sods), the source of the famous plaggen manure with its adhering sand (see also the preceding section).

Staring, a nineteenth-century agronomist and the first Dutch pedologist, points out in several of his publications that the existence of common heathlands and the shortage of manure retarded reclamation. In 1886 the Enclosure Law (*Markenwet*) permitted individual farmers to ask for the common lands to be divided (Lamberts, 1971, p. 257). However, large-scale reclamation of the heathlands only began after the First World War when fertilizers became readily available (Edelman, 1950, p. 32; Fig. 30). The enclosure and reclamation of the heathlands reduced the possibility of turf-cutting. At the same time, the higher yields of rye (Fig. 30) and the

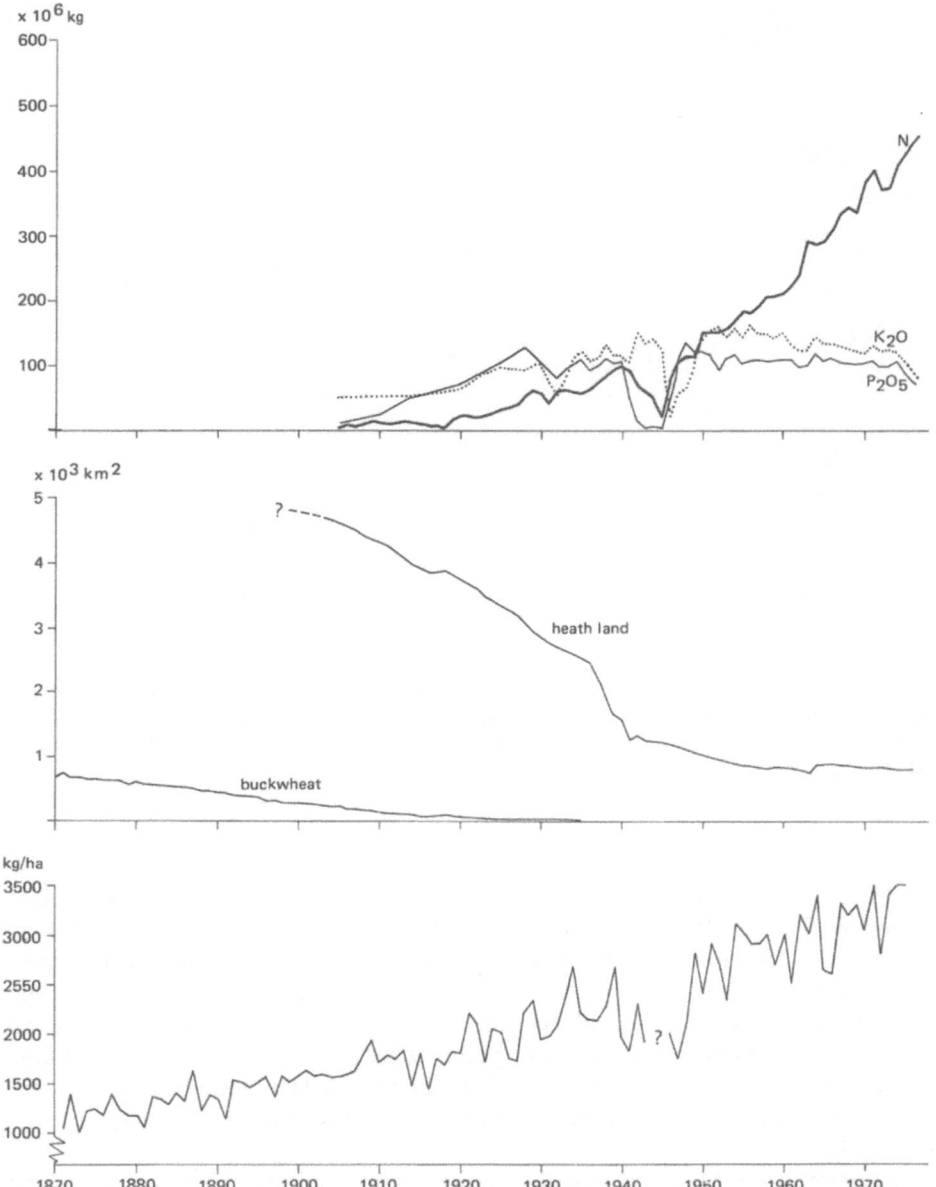

Fig. 30. Consumption of fertilizers (expressed as N, K_2O and P_2O_5), the disappearance of buckwheat, the decrease of heathland and the increase of the average yield of rye in the past hundred years in the Netherlands.

increase in the area of arable land enabled the farmers to use more straw as bedding material in the stables. This dung-impregnated straw did not contain sand (as did the plaggen) and therefore when spread on the fields it rotted down and did not raise the level of the arable soils. The improved fertility of the sandy soils is not only

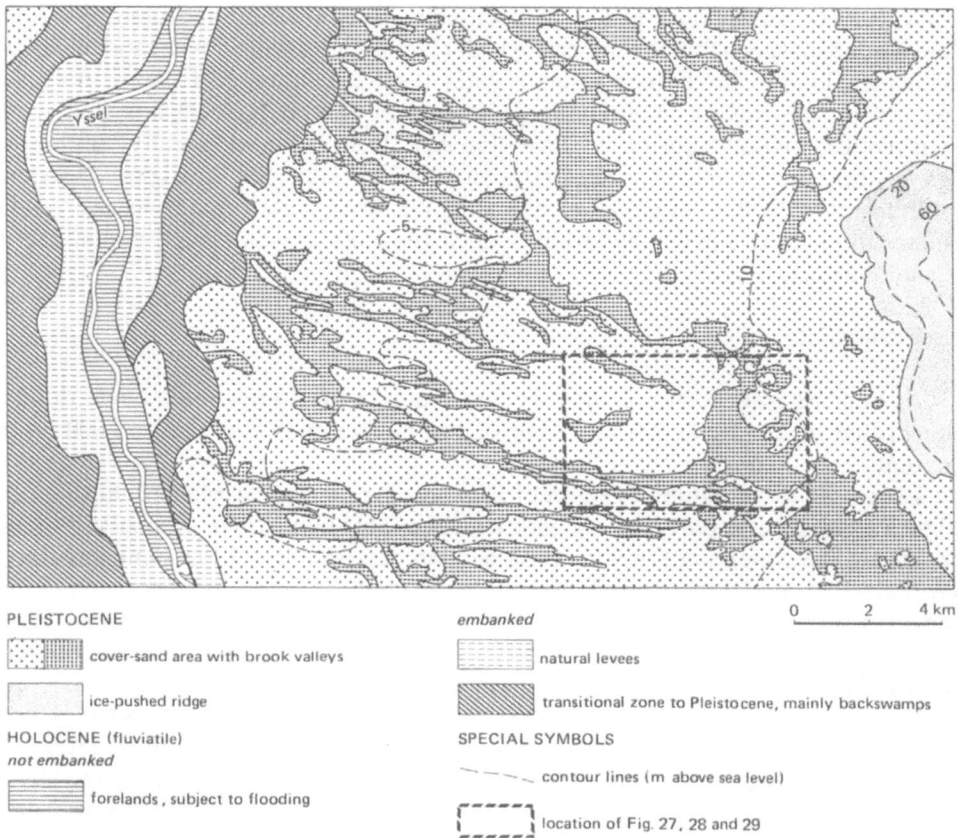

PLEISTOCENE

cover-sand area with brook valleys

ice-pushed ridge

HOLOCENE (fluviatile)
not embanked

forelands, subject to flooding

embanked

natural levees

transitional zone to Pleistocene, mainly backswamps

SPECIAL SYMBOLS

contour lines (m above sea level)

location of Fig. 27, 28 and 29

0   2   4 km

Fig. 31. Some geographical elements in the Pleistocene sandy district. The cover-sand area for which the Figures 27, 28 and 29 give detailed information, slopes gently from 15 m above sea level at the foot of the ice-pushed ridge in the east, to 5 m above sea level, where it dips under the fluviatile sediments of the river Yssel. Shallow valleys with Humaquepts drain this area westwards. For location see Figure 3.

shown by the higher rye yields (rye was until recently the typical crop on sandy soils in the Netherlands), but also by the disappearance of buckwheat, a crop grown on poor soils (Fig. 30).

Four of the soils found in the Pleistocene sandy district do not occur in the sample area (Fig. 28).

Soil P2 (p. 161) is a Haplaquod with an arable layer thickened by plaggen manure in the same way as the soil represented by the fourth mapping unit in the sample area. It is also developed from cover sand, but unlike the soils shown in Figure 28, it has glacial till in the subsoil, which is typical for the Pleistocene sandy district in the north of the Netherlands (Fig. 2).

Soil P3 (p. 165) is a non-hydromorphic humuspodzol (a Haplohumod) which is encountered on thick cover-sand deposits with deep ground-water tables. On these soils the last remnants of heathlands can be found, but usually they are forested, mainly with Scots pine.

Soils P4 (p. 169) and P5 (p. 173, in fact a subsoil) are also non-hydromorphic soils developed on sands that are mineralogically less poor than the cover sand and that, like the Humod, have deep ground-water levels. This type of parent material occurs on the preglacial fluviatile sediments in central Netherlands that were pushed into low hills during the Saale Ice Age. It is also found locally on Pleistocene Rhine and Meuse terraces (too small to be indicated in Figure 2) in the southeastern Netherlands. When the soils on these parent materials are sandy loam, the land use is usually arable; when the soils are coarser, the land use is usually forest.

Loess district

As appears from Figure 2, loess occupies only a small area in the Netherlands; according to the 1:200 000 Soil Map of the Netherlands it is only 63 000 ha in extent.

Loess can be found in the south of Limburg Province near Maastricht (Fig. 2). The loess district is part of the West European loess belt, which is bordered in the north by the cover-sand region. Both aeolian deposits are of the same age (late-Weichsel, cf. p. 11) and both blanket older formations. The transition zone to the north is rather narrow (Fig. 34), and is called *zandleem* in Belgium (Atlas, 1950-1972, mainly mapping units 26-29) and *Sandlösz* in the Federal Republic of Germany (Maas & Mückenhausen, 1970, mainly mapping unit 19).

Loess is a fine-grained material with 10-20% clay and less than 20% sand. Thick deposits are calcareous below 2.5-3 m depth (cf. the C2 horizon on p. 180).

Next to soils derived from loess, the Dutch loess district is characterized by a small area of soils developed from locally outcropping older sediments. For example, the upper Cretaceous chalk has given rise to a rendzina, soil L3, p. 186-189 (which occupies only 2% of an area of 12 760 ha in the loess district surveyed in 1967).

Moreover, the landscape of the loess district is characterized by (for Dutch circumstances) large differences in elevation over short distances. This is clear from the stereo photograph (Fig. 32) and from the contours on the geomorphological map (Fig. 33).

The characteristics of the loess district are quite different from those in other districts. It gives an outlandish impression to tourists (it is a well-known Dutch holiday resort), and for many a junior pedologist it is also the first confrontation with phenomena uncommon in the Netherlands. These phenomena are: no ditches and no ground water at shallow depth (apart from the small river valleys); a medium-textured material (the Pleistocene wind-blown loess) that is not of Holocene marine or fluviatile origin; dominantly brownish colours instead of the greyish colours of the hydromorphic polder soils; progressive soil formation, not only deep decalcification but also formation of a textural B horizon; the only authochthonous soils derived from chalk; the pronounced relief: dry valleys and plateaux, slopes with small man-made scarps (Dutch *graften*) and sunken roads leading from the valleys to the plateaux.

The part of the recent floodplain of the Meuse that is depicted in Figure 34, slopes from +50 m in the south tot +20 m in the north, a drop of 30 m over a distance of 80 km. The small river in the south of Figure 33, the Geul, discharges into the Meuse north of Maastricht and slopes from + 90 m in the east tot + 40 m in the west, a drop of 50 m over a distance of 17 km. The Geul has only few small tributaries, such as the one entering from the northeast; most of the valleys are dry, i.e. streamless (see Fig. 33). The dry valleys from the plateaux are generally elevated 60-100 m above the valley of the Geul.

The relief of plateaux, slopes, dry valleys and small river valleys originated in the early-Pleistocene. It was partly caused by the incision of the Meuse and its tributaries, and partly by snow-melt water eroding the present-day dry valleys. In the southwest of the area of Figure 34 the relief was mainly produced by the Pleistocene Meuse, which was rejuvenated by the uplift of the Ardennes Massif and dissected the late-Tertiary peneplain (Van den Broek & Van der Waals, 1967). This peneplain consisted mainly of Cretaceous chalk and local Tertiary sediments. The river partly cut into its own sediments, leaving terrace remnants on different levels. In the southwestern part several remnants of Meuse deposits can be distinghuised between 40 m above sea level (the recent floodplain) and about 190 m above sea level (the remnants of the peneplain, dating back to the Pliocene-Pleistocene boundary). On the hill in the south of Figure 32 a remnant of a Meuse terrace, most probably Tiglien, is present at the + 150 m contour.

Beginning in the Saale ice age and continuing in the Weichselien the area was blanketed by loess. In about one third of the loess district the loess is between 5 and 10 m thick, only rarely thicker; about half of the area has a thinner covering of loess and one fifth of the area has either Holocene alluvial deposits in the river valleys or older material outcropping on the slopes.

The loess-cover has smoothed the pre-existing relief, but not obliterated it: the dry valleys that are so typical of the Dutch loess district are not caused by Holocene man-induced erosion. There has been some accelerated soil erosion, but mostly sheet erosion. The Hapludalfs (soils with a textural B horizon, such as soil L1) on the plateaux have an A2 horizon below the plough layer; on the slopes and at the heads of the dry valleys the Ap horizon rests directly on the Bt horizon or is developed in the Bt horizon. Only locally, on steeper parts, the solum disappears and the C horizon is at the surface; rarely the calcareous material is exposed. In consequence of this erosion, most dry valleys have buried soils with a thick overburden of colluvium.

In the 1967 survey of the loess district, 10 150 ha of the 12 760 ha surveyed were found to be loess soils, and of these the non-eroded loess soils comprised 37%, the slightly eroded soils 30%, the soils with eroded solum 10%, and the colluvial soils 22%.

The steeper slopes of the valleys in the loess district have hardly any loess-covering and are characterized by small man-made scarps (Fig. 33). In the upper part of the narrow terraces the chalk is very near the surface, in the lower part it is covered with a mixture of loess and weathering products of the chalk, including flints. These scarps are very similar to the lynchets described by Orwin & Orwin, in the Appendix to their book entitled 'A note on the origin of Lynchets' (1967, p. 175-179); in the Netherlands they have been described by Breteler & Van den Broek (1968).

The sunken roads, called 'hollow roads' by the same authors, are a conspicuous phenomenon in the loess district. They lead from the farms in the valleys (Fig. 32) to the plateaux; the lower parts may be incised as deep as 5 m, the upper part 'surfaces' on the plateau. These roads are deepened tracks gradually made by carts

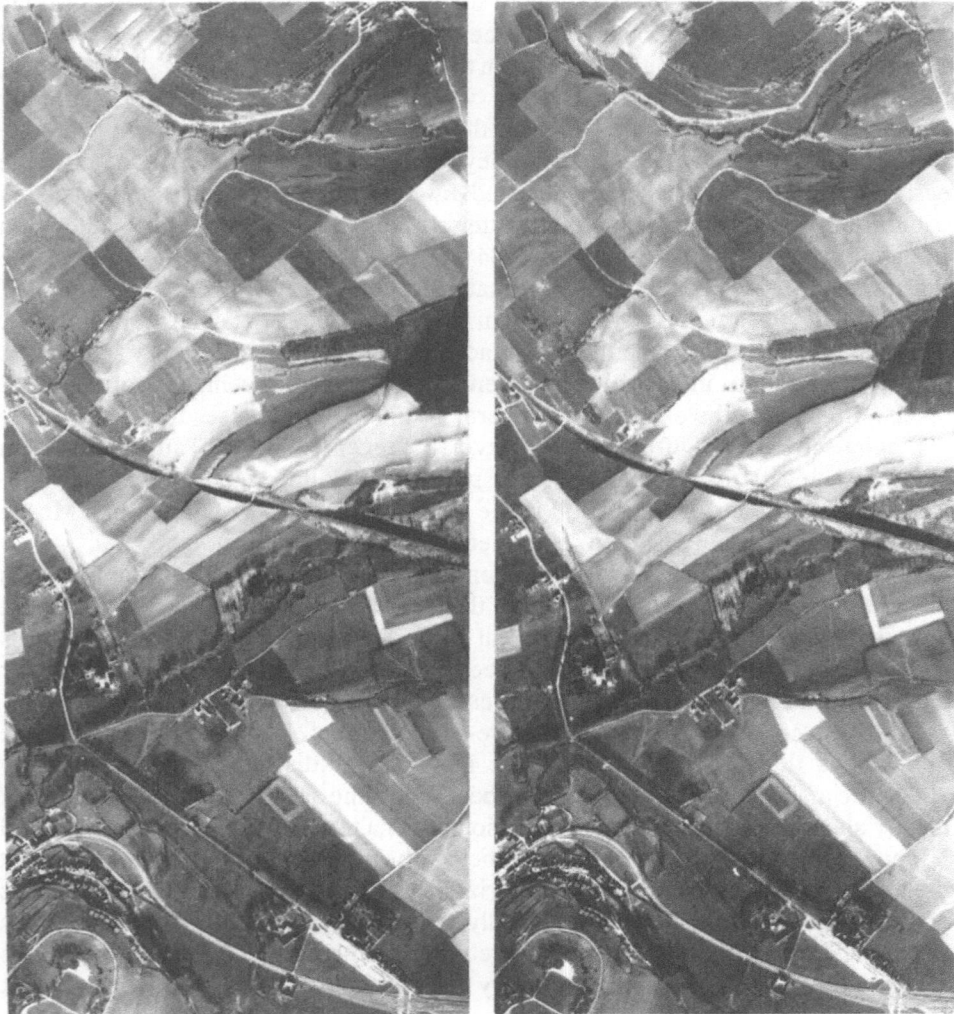

Fig. 32. Stereograph of a part of the loess district. In the north, where a reallocation scheme has been completed, traces of old roads are visible. The steeper parts of the slopes, with the lynchets (small scarps), are used as grassland, as are the valleys. Farms and villages are situated in the valleys because the ground-water level below the plateau surface is too deep. Conspicuous stereoscopic features are: the hill in the south; the plateaux in the north; the lynchets and the railway with cuttings and embankments. Same area as Figure 33.

and draught-animals and subsequent erosion of the originally unsurfaced roads. They often used the dry valleys as natural ramps to go uphill.

The loess district was settled very early; artefacts from the Palaeolithic have been found. Settlements dating from the time of the *Band-keramik* culture (4400-4000 B.C.) have been excavated (Atlas, 1963-1977, sheet VIII-1-C).

Roman settlement was rather dense. The loess district benefited from the *Pax Romana* because it was well within the pacified area: 200 km south of the northern boundary of the Roman Empire, which was at the Rhine. Many *villae* have been excavated and there are evidences of a Roman centuriation of the cultivated

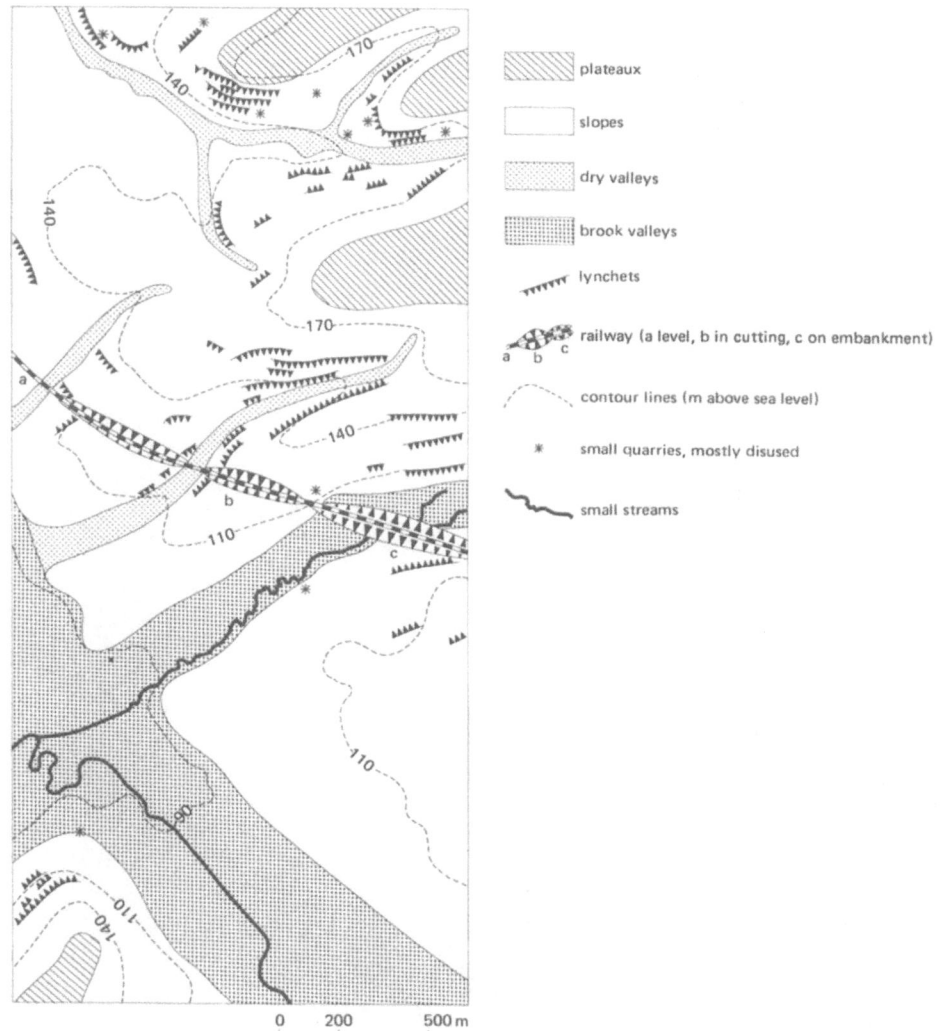

plateaux

slopes

dry valleys

brook valleys

lynchets

railway (a level, b in cutting, c on embankment)
a b c

contour lines (m above sea level)

* small quarries, mostly disused

small streams

0   200   500 m

Fig. 33. Simplified geomorphological map from a part of the loess district. The plateaux have thick loess deposits with Typic Hapludalfs (soil L1). On the steeper parts of the slopes lynchets are abundant and Lithic Rendolls (soil L3) have developed from outcropping Creataceous chalk. Locally Typic Eutrochrepts can be found in thick weathering products of the chalk (Sample 10a in Table 1). The dry valleys have thick colluvium derived from loess with Typic Udifluvents and the valleys with streams have thick alluvium sometimes overlying peat, with Typic Fluvaquents. For location see Figure 34.

ground (Lambert, 1971, p. 39). Maastricht was a Roman settlement before 50 A.D. at a ford in the Meuse, a site from which she derived her Roman name: *Mosae Trajectum.* There are remnants of the walls of a *castellum* and of *thermae.* Maastricht was a bishop's seat at the end of the fourth century and a *Palts* town (temporary seat of the emperor) in Carlovingian times. Heerlen, the large urban area 20 km northeast of Maastricht (Fig. 34), the Roman *Coriovallum,* also had *thermae* and when the Roman boundary was moved southwards at the end of the third century this town was fortified. The last Roman remains date from the end of the fourth century.

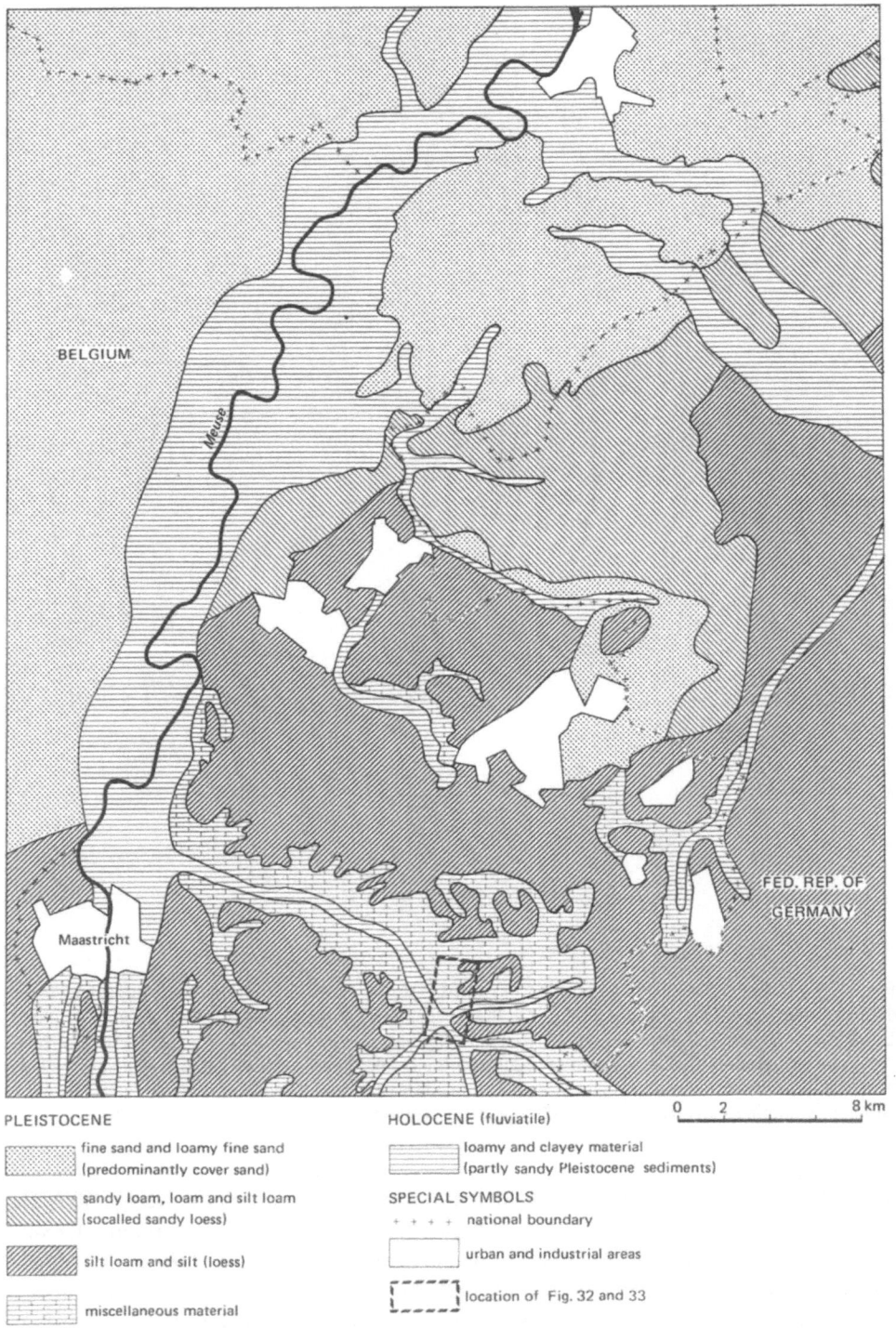

PLEISTOCENE

⬚ fine sand and loamy fine sand
(predominantly cover sand)

⬚ sandy loam, loam and silt loam
(socalled sandy loess)

⬚ silt loam and silt (loess)

⬚ miscellaneous material

HOLOCENE (fluviatile)

⬚ loamy and clayey material
(partly sandy Pleistocene sediments)

SPECIAL SYMBOLS

+ + + +   national boundary

⬚   urban and industrial areas

⌐ ⌐ ⌐ ⌐   location of Fig. 32 and 33

0   2   8 km

Fig. 34. Parent material and surface geology in the southern part of the Netherlands and the adjacent parts of Belgium and the Federal Republic of Germany. The miscellaneous material is found on the slopes of this dissected loess plateau where Upper Creataceous sediments outcrop (predominantly chalk); it consists of colluvial loess and weathering products of the chalk. For location see Figure 3.

As can be inferred from the presence of the Roman *villae,* much land must have been reclaimed for agriculture, but little is known from the following ages. Some settlements and their surrounding grounds were continuously used, others were abondoned. Most probably much of the land reverted to forest. Some of the names of the new settlements from the 11th-12th century have endings of *rode, rade* or *rath* from the verb *rooien* (= to root up, to grub up); other place names from that time end *holt* (= a wood). Certainly most of the land has been used for agriculture since the Middle Ages; today's forests are restricted to the steepest slopes, and heathland is only found in the small sandy area northeast of Heerlen (Fig. 34).

Three soils have been chosen to illustrate the soil conditions of this district. Soil L1 is a Typic Hapludalf, the central concept of a soil with a textural B horizon developed from loess (p. 178-181), it is the most common soil in the district. The other two soils are both rare. Soil L2 is a hydromorphic soil with a textural B horizon, which is only found locally on the northern boundary of the loess district, its classification in the different systems is somewhat controversial (p. 182). Soil L3 is only to be found on the slopes with outcropping chalk, it is a classic rendzina (p. 186-189).

Soils – discussion and classification

Each of the 32 coloured plates in this chapter is accompanied by the sections: General data, Discussion, Profile description, and Analytical data.

The first section has the subheadings: Classification, Location, Parent material, Topography and elevation, Drainage and ground water, Present land use, and Range in land use.

Under the sub-heading Classification the soils have been placed in ten taxonomies (lists of soil units, legends, and systems of soil classification):

1. *Eur.:* the Soil Map of Europe, scale 1:2 500 000 (Dudal et al., 1966, p. 3-7);
2. *World:* the Soil Map of the World, scale 1:5 000 000 (FAO, 1974);
3. *USA '38:* the Yearbook of Agriculture (USDA, 1938, p. 979-1001), with modification (Thorp & Smith, 1949);
4. *USA '75:* Soil Taxonomy (SSS, 1975);
5. *Ger.:* Committee on Soil Classification of the German Society of Soil Science (Mückenhausen et al., 1977);
6. *Eng. & Wales:* Soil Classification in the Soil Survey of England and Wales (Avery, 1973);
7. *France:* Committee on Soil Science and Soil Survey of France (CPCS, 1967);
8. *Neth. '50:* Provisional Soil Map of the Netherlands, scale 1:400 000 (Edelman, 1950);
9. *Neth. '61:* Soil Map of the Netherlands, scale 1:200 000 (Stichting voor Bodemkartering, 1961);
10. *Neth. '66:* the Dutch System of Soil Classification (De Bakker & Schelling, 1966) including the codes of the mapping units of the Soil Map of the Netherlands, scale 1:50 000.

The other subheadings provide information about soil-forming and environmental factors for each of the 32 soils for the benefit of readers unfamiliar with Dutch conditions.

In the section Discussion the place of the soils in these ten systems is discussed. The classification according to systems 1-7 is discussed in greater detail than the classification according to the three Dutch systems (8, 9 and 10) which are of interest only to readers with special interest in the Netherlands systems. No comments are made about the place of the soils in the Legend of the Soil Map of Belgium, scale 1:20 000 (Tavernier & Maréchal, 1958), mainly because its terminology leans heavily on the old USA system. Occasionally it has been found useful to refer to other systems, such as Kubiëna's (1953), FitzPatrick's interesting system (1971), and some East European systems. The Canadian system (CSSC, 1970; Clayton et al., 1977), the Australian system (Northcote, 1971) and the South African system (MacVicar et al., 1977) are not discussed.

The notations assigned to the soil horizons are also discussed, because the symbols accompanying the plates were printed for the original Dutch version of this book (De Bakker & Edelman-Vlam, 1976) and they are those used in the Soil Survey Institute of the Netherlands (De Bakker & Schelling, 1966, p. 62-64).

Deviations from internationally know systems are explained and comments on differences between foreign systems are made.

The Discussion sections end with some brief comments on the analytical data that accentuate typical aspects of the soil in question.

The remaining two sections comprise the profile description and some analytical data. For the former, soil texture and soil structure terminology are according to Soil Taxonomy (SSS, 1975, p. 469-471 and p. 474-477), and the colours of the moist soil are described according to the Munsell code.

The analytical data include particle-size analysis and some chemical data; in Appendix 2 the analytical methods used are briefly discussed.

Soils of the marine district: M1

General data

Classification

Eur.	Alluvial soil (p. 4 and 75-77)
World	Calcaric Fluvisol (p. 33), fine textured over medium textured, level (p. 5), saline (p. 7)
USA '38	Bog soil, Coastal Marshland (p. 1000 and 1132)
USA '75	Fine-clayey over loamy, illitic (calcareous), mesic Hydraquent (p. 185)
Ger.	Vorland-Seemarsch (p. 158)
Eng. & Wales	Unripened gley soil
France	Sol minéral brut d'apport alluvial (p. 18)
Neth. '50	Neither on the map nor in the book
Neth. '61	Sea clay foreland soil, with a vegetation, more than 80 cm clay over sand; mapping unit 4
Neth. '66	'Slik' vague soil (p. 157 and 194); mapping unit MOo05A

Location. Zeeland Province, on a coastal marsh alongside the south bank of the river Scheldt, about 20 km southeast of Flushing (see Fig. 1).
Parent material. Very recent marine sediment, with continuing accretion.
Topography and elevation. Level; about mean high tide.
Drainage and ground water. Drained by a system of dendritic, narrow natural tidal creeks, by which only the external (tidal) water is removed, there is no internal drainage as the unripened mud has a very low permeability; consequently the ground-water level is always at the surface: the soil is a two-phase system, viz. soil and water, no air.
Present land use. Salt marsh with cord-grass (*Spartina × townsendii*) dominant, some marsh samphire (*Salicornia spp*) and sea aster (*Aster tripolium*).
Range in land use. Tidal flats silted to a higher level have a less frequency and height of the tidal flooding, and more variety in plant cover. They are used for extensive grazing by sheep. After embankment, drainage and oxidation they change into soils like profile M6.

Discussion

According to the tables in *Soils and Men* (USDA, 1938, p. 993-1001) it seems evident that all the Dutch coastal marshland soils, whether embanked or not, should be placed in the class of Alluvial Soils. However, on closer examination, it appears that alluvial soils have to 'occur. . . along streams' (p. 1133). Furthermore on p. 1128 it states that the great soil group Bog soils is 'represented by two general groups – peat and muck, and coastal marshland', but when reading the description of the latter group (p. 1132: 'generally the surface materials consist of . . . coarse, fibrous and matted peat') this does not cover Dutch coastal marshland. It must be concluded that there is no place for these soils in the old USA system.

On the Soil Map of Europe all recent fluviatile and marine soils are classified as Alluvial Soils, whether ripened or not. In the French system and on the Soil Map of the World no reference is made to soft consistence and high water content, which are diagnostic features of this soil grouping. Only the new USA system and the system of England and Wales have definite categories for non-ripened soils.

All soils on the forelands of tidal areas have to be classified as *Vorland-Seemarsche* (or *Brackmarsche,* or *Flussmarsche*) in the official German system (Mückenhausen et al., 1977, p. 158), but not all these soils are unripened. Following the meeting of Commissions V and VI on hydromorphic soils in Stuttgart in 1971, there was an excursion in the Federal Republic of Germany. The soils of the marine district of western Schleswig-Holstein seen on that occasion are worth considering in connection with this discussion (Hugenroth, 1971, p. 38-58). Profiles 1a and 1b of this German sequence are comparable to this soil and were named *Schlickwatt,* whereas Profile 2 was called *Salzmarsch*; this latter soil was partly ripened ('This soil is already aerated to a depth of 70-80 cm', Hugenroth, 1971, p. 38). Obviously in that area the *Vorland-Seemarsche* are subdivided into *Schlickwatte* and *Salz-marsché* based on the depth of ripening.

The blue soft mud is an example of a G horizon in the old USA system of horizon nomenclature (SSS, 1951, p. 181), in the Dutch system (De Bakker & Schelling, 1966, p. 63) and in the ISSS system (1967); of a Gr horizon in the German system (Arbeitsgemeinschaft Bodenkunde, 1971, p. 30); of a CG horizon in the French (Jamagne, 1967, p. 38) and British systems (Hodgson, 1974, p. 81), and of a Cr horizon in the revised ISSS system as published in the legend of the Soil Map of the World (FAO, 1974, p. 23). In the USA system the distinction between the old Cg and G horizon has been dropped, and both have to be coded as Cg (SSS, 1975, p. 461 and 462), as has been used in profile description No. 37 in the 7th Approximation (SSS, 1960, p. 119), on the other hand the Sulfaquent in Soil Taxonomy (SSS, 1975, pedon 43, p. 569) has no horizon designations.

This soil-to-be is flooded twice daily by sea water and this is reflected in the analyses. The percentage of exchangeable sodium and magnesium is high, as is the pH. Because the sulphur content is too low and the carbonate content too high, it is not sulfidic as defined in the USA system (SSS, 1975, p. 63) or in the legend of the Soil Map of the World (FAO, 1974, p. 31) and thus there is no possibility of this material turning into a sulfuric horizon on oxidation (SSS, 1975, p. 47; FAO, 1974, p. 27).

The gradual change of the site from a bare tidal sand flat into a salting with *Spartina* is also reflected in the analyses: during the upward growth of the foreland by accretion both clay content and organic matter content are increased. The organic matter: clay ratio is about 0.1 and the C-N ratio is relatively low; both are typical for unripened marine sediments (SSS, 1960, p. 119; Hugenroth, 1971, p. 44 and 46; Brümmer, 1968, p. 294). C-N ratios of fresh-water tidal sediments are much higher and drop from 20 to 10 following embankment and drainage (Zonneveld, 1960, p. 86). A possible explanation for this difference might be the relatively high amount of living organisms in the salt marshes compared with fresh-water tidal areas – 'the salt marsh, a vast pantry on the ocean shore' (Hitchcock, 1972, p. 738). This kind of organic matter has a much higher content of nitrogen than normal plant remains.

Profile description

AG	0-10/20 cm	Grey (5Y5/1) clay; slightly aerated; many roots of *Spartina*; few, faint, medium, yellowish brown (10YR5/4) mottles; apedal, with many coarse pores; consistency terminology not applicable for this nonripened, slushy material, which feels like soft soap, the site is only passable via the *Spartina* tussocks; clear, smooth boundary.
G1	10/20-80 cm	Dark grey (10GY4/1) clay loam grading gradually into loam with depth; no brown mottles, few, faint, medium mottles which are less bluish than the matrix colour; apedal, non-ripened, slushy material.
G2	80-100 cm	Dark grey (7.5Y4.5/1) fine sandy loam; no brown mottles; apedal, because of the higher sand content less slushy than the overlying horizon; clear, smooth boundary.
DG	100-120 cm plus	Grey (5Y5/1) fine sand; single grain; non-sticky, non-plastic, loose.

Analytical data

Hori-zon	Depth (cm)	Organic fraction				Particle-size distribution of mineral fraction (% of fine earth)						Field moisture (% of dry soil)
		O.m. (%)	C(%)	N(%)	C/N	< 2	2-16	16-50	50-105	105-150	> 150 µm	
AG	0-9	6.8	3.9	0.35	11.1	54	19	20	3.5	2	1.5	119.7
G1	43-55	4.4	2.3	0.20	11.5	36	13	24	19	6	2	74.4
G2	80-90	1.8	n.d.	n.d.	n.d.	16	5	15	20	29	15	n.d.

Horizon	CaCO3 (%)	pH-KCl	Extractable cations (% of sum)					meq/100 g			Ca/Mg	K-fix (%)
			Na	K	Mg	Ca	H	C.E.C.	CaO	SO4		
AG	17.9	8.4	49.5	8.2	11.2	29.5	0.6	32.9	349	15	2.6	n.d.
G1	18.5	8.6	31.0	7.8	19.4	41.4	0.7	31.9	374	36	2.1	n.d.
G2	13.7	8.7	n.d.	n.d.	n.d.	n.d.	n.d.	n.d.	n.d.	n.d.	n.d.	2

Soils of the marine district: M2

General data

Classification

Eur.	Alluvial soil (p. 4 and 75-77)
World	Calcaric Fluvisol (p. 33), fine textured over medium textured, level (p. 5)
USA '38	Bog soil, Coastal Marshland (p. 1000 and 1132)
USA '75	Fine-clayey, illitic (calcareous), cracked, mesic Hydric (?) Fluvaquent (p. 182 and 183)
Ger.	Kalkhaltige Seemarsch (p. 158)
Eng. & Wales	Calcareous alluvial gley soil
France	Sol minéral brut d'apport alluvial (p. 18) or sol d'apport alluvial hydromorphe (p. 25)
Neth. '50	Zuiderzee-bottom clay soil; mapping unit 16 (p. 150-153)
Neth. '61	Calcareous soil of the Zuiderzee-bottom, shallowly humuspoor, homogeneous down to more than 50 cm, clay; mapping unit 50
Neth. '66	'Nes' vague soil (p. 160 and 195); mapping unit Mo80Ap

Location. Oostelijk Flevoland Polder (see Fig. 1), this is the second youngest in Lake Yssel, the enclosing dike in this shallow lake was closed on 13 September 1956 and on 29 June 1957 the polder floor became dry.

Parent material. Very recent marine sediment.

Topography and elevation. Level; about 4 m below sea level.

Drainage and ground water. When sampled (spring 1963) this part of the polder was still being reclaimed and besides the ditches round the fields (300 × 1000 m) only the initial drainage system was present: the temporary field drains will in due course be replaced by tile drains; the ground-water level is already below the observed depth of the pit, but the subsoil is not yet completely ripened. The water level in the ditches is maintained at 5.20 m below the sea level.

Present land use. Ley pasture.

Range in land use. Dominantly arable land with the possibility of the same wide crop rotation as soil M6, but because the polders are embanked and reclaimed by the government, land use is planned; deciduous trees are planted on this field, mainly poplars.

Discussion

This very recent soil from a polder on the floor of the former Zuyder Zee, now Lake Yssel, still has a rather soft, partly ripened subsoil, thus being in an intermediate stage of ripening between soils such as the previous soft foreland soil and the completely ripened polder soil M6.

There is no place for such an intergrade in any of the systems discussed. In the new USA system it is an intergrade between a Fluvaquent and a Hydraquent, but such a subgroup has not been defined. The rules of nomenclature permit the use of the adjective hydric, so, such a new subgroup might be called a Hydric Fluvaquent. In the same system there is the possibility of using the modifier 'cracked' to designate families of Fluvaquents (SSS, 1975, p. 389). During the ripening process (Pons & Zonneveld, 1965, Fig. 7) soils of medium to fine texture, directly emban-ked from an unripened stage, form big prisms separated by desiccation cracks. Partly ripened soils and/or medium to coarse textured soils do not develop cracks upon embankment and subsequent drainage. So cracked Fluvaquents are very typical for the fine-textured soils of the new polders, which are being reclaimed by embanking parts of the shallow former Zuyder Zee. These cracks persist for a long time, and have the important effect of reducing the tile drainage required, whereas non-cracked or less cracked soils of the same texture need closely spaced tile drains.

In the French system the question arises about where to put the boundary between the *sols bruts* and *sols peu évolués*. It is known that this soil has been ploughed only recently. Stratification starts directly under the plough layer, where the only soil formation has been oxidation and the formation of the blocky structu-res in the C21g, cracks and big prisms in the C22g. The organic matter present amounts to more than traces (CPCS, 1967, p. 21), but it is the initial content inherited from the parent material (see previous soil); in time the content will be lowered by oxidation to a level similar to that in soil M6. The definitions do not solve this dilemma; perhaps the fact that the soil is used as arable land is justifica-tion for classifying the soil as a *sol peu évolué*.

The subsurface horizons produce no difficulties in the Dutch system of soil horizon nomenclature; they are clearly C horizons. In most international systems there is a discrepancy between the presence of soil structure which justifies a B horizon, and the presence of rock structures (such as stratification) which are diagnostic for C horizons.

The DG horizon (2Cr according to FAO, 1974) is formed from the Pleistocene deposits, abraded by the waters of Lake Flevo, the predecessor of the Zuyder Zee in Roman times (cf. p. 23).

The circumstances of sedimentation in this former shallow bay were very quiet, separated as it was by the Wadden Sea, and the Frisian Islands from the North Sea by a distance of more than one hundred km. This is reflected in the texture: a medium amount of clay and a very small quantity of sand, locally there are fine-silty soils that have a loess-like texture.

The amount of organic matter in the subsoil is large with a high C-N ratio. Both organic matter and C-N ratio decrease upwards towards the surface where calcium carbonate content is greatest. The explanation of these characteristics is straight-forward. During the development of the Roman Lake Flevo into Zuyder Zee the first sediments were strongly unfluenced by peat erosion from the margins of the lake. As the influence of the sea grew stronger, the inorganic component and amount of sea shells became greater.

The decreasing percentage of calcium and the increasing percentages of magne-sium and sodium from the surface point to the fact that chemical ripening is still in progress. Physical ripening is not yet completed either. The amount of moisture in the subsoil and its slushy character are typical of unripened material. As ripening penetrates deeper into the sedimentary material, the cracks will deepen and the 'soil' become firmer.

Profile description

Ap	0-20/25 cm	Dark grey (2.5Y4/1) silty clay loam, with some clods of lighter (both in colour and texture) material; no mottles; moderate, fine blocky structure; very friable, slightly hard; abrupt, smooth boundary.
C21g	20/25-32/42 cm	Dark greyish brown (2.5Y4/2) silty clay; few, faint, fine yellowish brown mottles (10YR5/5) in root channels and on ped faces; cracked to strong, blocky elements; many shell fragments; slightly sticky, plastic, friable; abrupt, smooth boundary.
C22g	32/42-62 cm	Olive grey (5Y4/2) silty clay loam; common, distinct fine brown mottles inside the prisms; cracked to strong very coarse prism which are stratified; some of the prisms are cut through, the insides have some 'bluish' mottles (N4 - 5GY4/1) proving the soil is not yet fully oxidized; slightly sticky, plastic, friable; abrupt, smooth boundary.
C23g	62-82 cm	Very dark brown (10YR2.5/1.5) silty clay loam to silty clay; few, faint, fine brown mottles; apedal slightly slushy material, can be squeezed with difficulty through the fingers; stratified; abrupt, smooth boundary.
C2G	82-115 cm	Black (2.5Y2/1) silt loam mixed with allochthonous eroded disintegrated peat; few mottles; not yet ripened apedal, slushy material, can be squeezed between the fingers; abrupt, smooth boundary.
DG	115-120 cm plus	Grey (5Y5/1) fine sand; some mottles in the upper part, reduced in the lower part; resedimentated, loose, non-calcareous, Pleistocene sand.

Analytical data

Horizon	Depth (cm)	Organic fraction				Particle-size distribution of mineral fraction (% of fine earth)					Field moisture (% of dry soil)
		O.m. (%)	C(%)	N(%)	C/N	< 2	2-16	16-50	50-105	> 105 μm	
Ap	0-20	3.1	1.6	0.14	11.4	34	21.5	36	7.5	2	n.d.
C21g	25-32	3.7	n.d.	n.d.	n.d.	47	28.5	17.5	3.5	3.5	n.d.
C22g	42-62	6.2	3.1	0.23	13.5	31	45	20.5	1.0	2.5	64.0
C23g	62-82	10.2	6.1	0.26	23.5	27	35.5	36	0.5	1	89.9

Horizon	CaCO₃ (%)	pH-KCl	Extractable cations (% of sum)				meq/100 g			Ca/Mg
			Na	K	Mg	Ca	C.E.C.	CaO	SO₄	
Ap	10.7	7.3	1.7	3.9	13.8	80.6	23.2	n.d.	n.d.	5.7
C21g	10.1	7.5	0.6	5.2	18.5	75.7	25.3	202	30	4.1
C22g	9.6	7.6	2.8	4.1	24.6	68.5	25.3	192	41	2.8
C23g	8.4	7.6	9.0	2.8	29.3	58.9	32.9	168	30	2.0

0	
	Ap
25	
	C21g
32	
	C22g
62	
	C23g
82	
	C2G
115	
	DG

Soils of the marine district: M3

General data

Classification

Eur.	Alluvial soil (p. 4 and 75-77)
World	Eutric Fluvisol (p. 32), fine textured, level (p. 5)
USA '38	Bog soil, Coastal Marshland (p. 1000 and 1132)
USA '75	Fine-clayey, illitic (nonacid), mesic Typic Fluvaquent (p. 182 and 183)
Ger.	Knick-Seemarsch (p. 158)
Eng. & Wales	Pelo-(vertic) alluvial gley soil
France	Sol d'apport alluvial hydromorphe (p. 25)
Neth. '50	Sticky clay soil; mapping unit 7 (p. 133-137)
Neth. '61	Non-calcareous, homogeneous, humus-poor and shallowly humose, young sea clay soil, heavy clay with unfavourable properties; mapping unit 34
Neth. '66	'Polder' vague soil (p. 160 and 195); mapping unit kMn43C

Location. Friesland Province (see Fig. 1), in the centre of the Frisian dairy district.
Parent material. Marine sediment, deposited about 1500 years ago.
Topography and elevation. Level; about 0.5 m below sea level.
Drainage and ground water. The field is not tile drained, but has a system of open drains (furrows), discharging into ditches with an irregular pattern (being partly relic tidal creeks); the ground-water level fluctuates between 30 cm in winter and 130 cm in summer.
Present land use. Permanent grassland, perhaps more than a thousand years old.
Range in land use. Exclusively grassland, with alternating use as pasture and meadow (both hay and silage grass).

Discussion

In the marine district of northwestern Germany and in the north of the Netherlands there are special marine soils called *Knick* (Ger.) (Mückenhausen et al., 1977, p. 158) and *knik* or *knip* (Neth.) (Edelman, 1950, p. 125, 135 and 136). They are non-calcareous, sticky, fine-textured soils, which traditionally are separated from similar soils by their lower permeability and higher density. There has been (Müller, 1954; Brümmer, 1968; Mückenhausen et al., 1977, p. 162) much discussion about their genesis, especially about the influence of the Ca-Mg ratio (is this ratio acompanying, diagnostic or defining the *Knick*-phenomena?).

The micromorphology of these soils is characterized by 'deformation of argillans caused by churning' (Jongerius, 1970, Fig. 6). Both the micromorphological and chemical characteristics are too weakly expressed to satisfy the definitions of the natric horizon (SSS, 1975, p. 28) and of the Natric B horizon (FAO, 1974, p. 26), although these soils were earlier classified as intergrades between Solonetz and Humic-Glei soils by Bennema et al. (1953, p. 48).

Concerning their place in the old USA system and in the legend of the soil maps of Europe and of the World nothing needs to be added to the discussion accompanying soil M1. In the system of England and Wales no distinction is made at subgroup level between soils from marine and fluviatile alluvium, so this soil has the same name as the fine-textured fluviatile soil F1.

If the blocky subsurface horizon is a (B) in the French system, this soil has to be classified at subgroup level as a *sol hydromorphe peu humifère à gley profond* (CPCS, 1967, p. 75), within the class of the *sols hydromorphes*. This difficulty in classification was also experienced by Monsieur Servant who surveyed polders on the French coast of the Strait of Dover: '. . . accordingly one could speak of a weakly developed soil of marine origin or of a mineral hydromorphic gley soil. . .' (Servant, 1973, p. 40), meaning if there is a (B) it is a *sol peu évolué* and if there is a C it is a *sol hydromorphe*.

The subsoil has a pronounced structure and shows neither fine laminations nor big textural differences, however there are two buried A1 horizons, and there are doubts, as with soil F1 (p. 91) whether to designate it a Bg, a (B)g or a Bwg horizon, or, as in the Dutch system a Cg horizon. There are no difficulties either in the German system in which it would be a Go horizon (compare with soil 58 in Mückenhausen et al., 1977, p. 292 and 293, which is practically identical).

The difference between the non-calcareous upper horizons and the deeper calcareous subsoil is indicated by using C1 and C2 respectively.

In common with the German *Knickböden* this soil also has a very low Ca-Mg ratio. Müller (in: Hugenroth, 1971, p. 68) uses this ratio as one of the criteria to subdivide soils in the coastal marsh. The low Ca-Mg ratios are indeed characteristic for *Knickböden,* but are an accompanying property rather than a diagnostic criterion. Moreover, there are non-*knick* soils that also have rather low Ca-Mg ratios, e.g. soils with brackish ground water and soils that were inundated during the last world war. Shortly after inundation, both sodium and magnesium percentages were high. However, sodium is more readily exchangeable by calcium than is magnesium, so the relatively high percentage of magnesium in such cases is a last trace of war damage (Van der Molen, 1957, Table 49, p. 104).

As in all old grasslands on fine-textured soils, organic matter is high in amount in the sod, and in the subsoil it is still enough to qualify for the 'fluvic' characteristics of this Fluvaquent, i.e. more than 0.2% organic carbon or more than 0.35% organic matter (SSS, 1975, p. 182).

Profile description

A1g	0-12 cm	Dark grey (2.5Y3.5/1) silty clay; few mottles; in the upper few cm's moderate very fine subangular to medium crumb structure, in the lower part moderate coarse compound prisms breaking to weak very fine to fine subangular blocky structure; friable, slightly hard; abrupt, smooth boundary.
ACg	12-30 cm	Grey (2.5Y4.5/1) silty clay; common, distinct fine brown mottles; moderate to strong very coarse prisms, breaking to moderate medium subangular blocky structure; firm, slightly hard; clear, smooth boundary.
C11g	30-56 cm	Grey (5Y4.5/1), silty clay; few to common, faint, fine brown mottles; weak to moderate very coarse compound prismatic structure breaking to moderate coarse angular blocky structure; firm to very firm, hard; clear, smooth boundary.
C12gb	56-90 cm	Grey (7.5Y5/1), silty clay; common, faint, fine brown mottles; the upper part is a buried A1; strong coarse angular blocky to medium prismatic structure; very firm, hard; clear, smooth boundary.
C2gb	90-120 cm	Grey (7.5Y5/1), silty clay, just below 120 cm depth, grading into stratified sandy loam to loam; few, faint, fine brown mottles; calcareous; the upper part is a buried A1; weak to moderate, medium to coarse angular blocky structure; firm, hard.

Analytical data

Horizon	Depth (cm)	Organic fraction				Particle-size distribution of mineral fraction (% of fine earth)			
		O.m. (%)	C(%)	N(%)	C/N	< 2	2-16	16-50	> 50 μm
A1g	0-8	14.7	7.7	0.75	10.3	41	18	34	8
ACg	12-20	6.8	3.8	0.40	9.5	41	23	31	6
C11g	30-40	1.8	n.d.	n.d.	n.d.	51	18	30	2.2
C2gb	100-110	0.7	n.d.	n.d.	n.d.	43	29	25	2.3

Horizon	CaCO₃ (%)	pH-KCl	Extractable cations (% of sum)					C.E.C. (meq./ 100 g)	Ca/Mg	K-fix. (%)
			Na	K	Mg	Ca	H			
A1g	0.0	5.2	n.d.	n.d.	n.d.	n.d.	n.d.	n.d.	n.d.	n.d.
ACg	0.0	5.2	n.d.	n.d.	n.d.	n.d.	n.d.	n.d.	n.d.	n.d.
C11g	0.3	6.1	1.7	3.4	32.6	58.9	3.4	23.6	1.8	24
C2gb	3.3	6.9	1.6	3.6	27.4	65.2	2.0	24.7	2.4	29

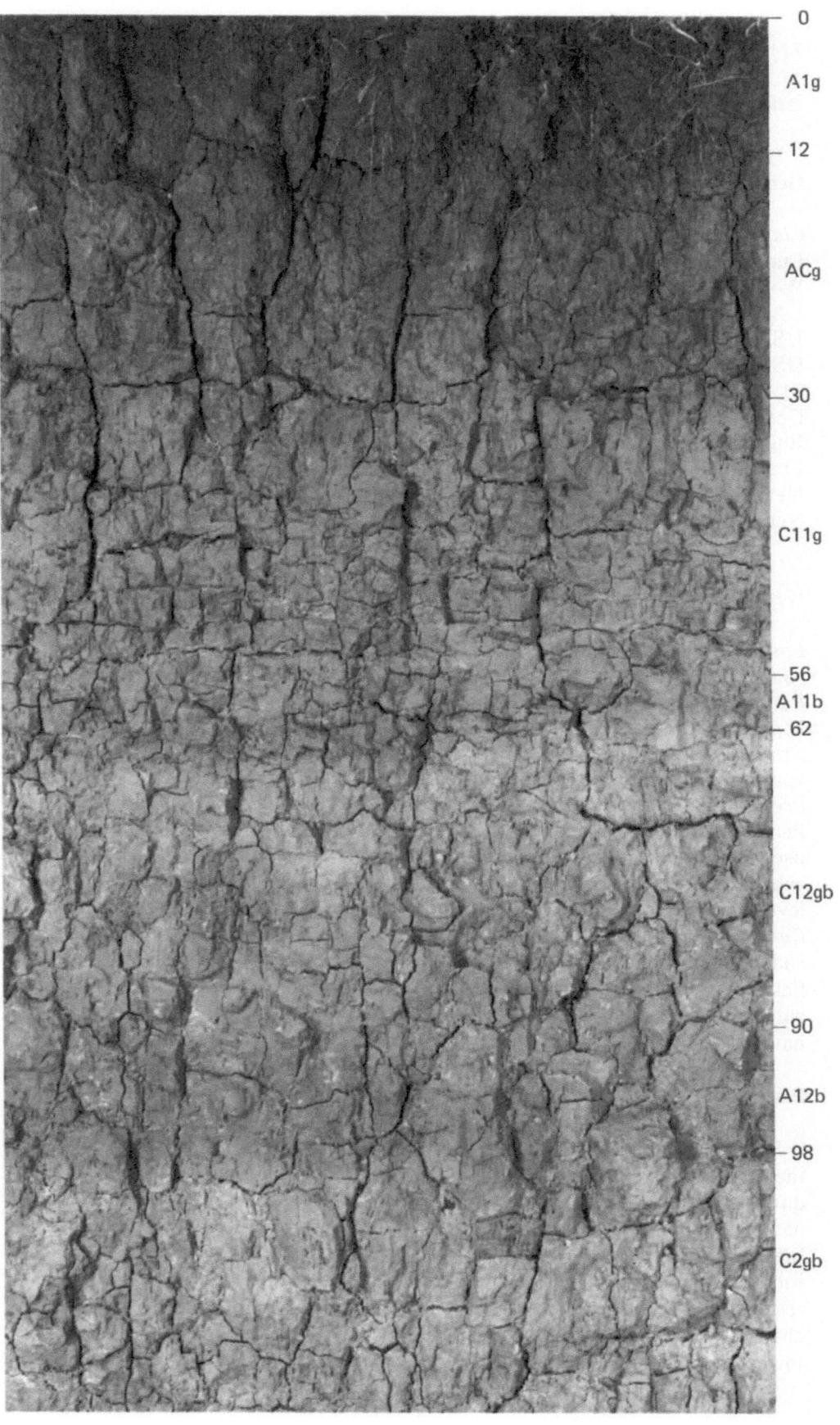

0	
	A1g
12	
	ACg
30	
	C11g
56	
	A11b
62	
	C12gb
90	
	A12b
98	
	C2gb

Soils of the marine district: M4

General data

Classification

Eur.	Alluvial soil (p. 4 and 75-77)
World	Calcaric Fluvisol (p. 33), medium textured over coarse textured, level (p. 5)
USA '38	Bog soil, Coastal Marshland (p. 1000 and 1132)
USA '75	Fine-loamy over sandy, mixed (calcareous), mesic Fluvaquent (p. 182 and 183), proposed to be called Psammic
Ger.	Kalkhaltige Seemarsch (p. 158)
Eng. & Wales	Calcareous sandy gley soil
France	Sol d'apport alluvial hydromorphe (p. 25)
Neth. '50	Young marshy sandy silt soil; mapping unit 14 (p. 111)
Neth. '61	Calcareous, predominantly clay-poor sand beginning within 50 cm, humus-poor and shallowly moderately humose, young sea clay soil, complex of clayey sand to clay; mapping unit 11
Neth. '66	'Vlak' vague soil (p. 157, 158 and 194); mapping unit kZn40A

Location. Zeeland Province, some 20 km southeast of Flushing (see Fig. 1), on the south bank of the Scheldt estuary, in the Koninginnepolder (embanked in 1893).
Parent material. Recent marine sediment; thin loam overlying fine sand.
Topography and elevation. Level; about 1.5 m above sea level.
Drainage and ground water. Not tile drained, no furrows, ditches spaced 50-150 m apart; ground-water level fluctuates between 1 and 2 m.
Present land use. Arable land: Lucerne (*Medicago sativa*), after malting barley. Part of the field is being excavated. About 1 m of the sandy subsoil is removed to be used for road building, the fine-textured topsoil is replaced on the remaining sandy subsoil and the field reclaimed and used for grassland with a higher ground-water level than in the original soil. The soil was sampled from the wall of the excavation.
Range in land use. These soils occupy small areas surrounded by deeper loamy soils, therefore the range in land use is suited to these soils (see soil M6). If whole fields have shallow sand, the farmer takes into account that these soils are liable to suffer from drought because of their inability to retain water; therefore such fields have less sugar beet and more spring grains in the crop rotation.

Discussion

Soil M4 is a shallow loam over sand, like soil F2 from the fluviatile district, therefore the discussion about the classification of these alluvial soils in the different systems is similar, with one exception. Only the German system discriminates between marine, *Marsche* and fluviatile soils, *Vegas* or *Auenböden*. Therefore, this marine soil must be a *Marsch* and because the soil is calcareous, it is a *kalkhaltige Seemarsch*. Differences in 'particle-size distribution of the horizons, respectively layers of coastal marsh soils' (called 'texture profile' by Green, 1968, chapter IV; and by Northcote, 1971, p. 29) are used as differentiating criteria at the lowest level of classification which is called *Form* in the German system (Mückenhausen et al., 1977, p. 158).

In 1866 Staring (one of the Dutch founding fathers of pedology, who called himself an 'earth scientist') published a kind of soil map of the Netherlands, scale 1:200 000. Since that time, a subdivision has existed in the Netherlands *zandgronden* (coarse-textured soils) and *kleigronden* (medium- and fine-textured soils). This traditional subdivision has been adopted in our new system. The difference between the two classes can be roughly indicated by: sand shallower, or deeper than 40 cm (for the complete definition see De Bakker & Schelling, 1966, p. 179). An analogous distinction is made by Soil Taxonomy, e.g. Psammaquents versus Fluvaquents, but in this system the depth of 25 cm is the boundary between both Great Groups (SSS, 1975, p. 185 and 182). Thus soil M4 is still a Fluvaquent (apart from the organic-matter level in the subsoil). The adjective Psammic instead of Typic is used because of the shallow depth to the sand.

The system of England and Wales distinguishes groups of alluvial gley soils and sandy gley soils; the boundary is defined as follows: 'For inclusion in sandy groups or subgroups, at least half the upper 80 cm of mineral soil. . . must be sand or loamy sand.' (Avery, 1973, p. 334). This definition has been taken from our system (De Bakker & Schelling, 1966, p. 179), in both systems the soil belongs to the sandy subgroup.

The loamy cover of the former sand flat is 35 cm deep, 25 cm of it being the plough layer (Ap); the remainder is designated ACg. On the photograph the stratification in the ACg is visible (see also the profile description). The textural change to the sandy subsoil is very abrupt; this grey mottled subsoil is labelled Dg, in other systems: 2Cg (FAO, 1974, p. 22), also Dg in the old USA system (SSS, 1951, p. 180 and 181), IICg (SSS, 1975, p. 461 and 462), Go (Arbeitsgemeinschaft Bodenkunde, 1971, p. 30), 2Cg (Hodgson, 1974, p. 72 and 80), and IICG (CPCS, 1967, p. 12 and 13).

In the first section of this discussion the similarity between this profile and soil F2 has been mentioned. However, chemically they are unlike, as is determined by the difference in potassium fixation between marine M4 (13%) and fluviatile F2 (33%) soils.

Sandy subsoils in the marine area are always fine-grained (there are no shingle beaches on the Dutch coast). In contrast sandy subsoils in the riverine area are coarse. For example, the subsoils of soils M4 and F2 have nearly the same amount of sand, but the sea sand has only 0.5% over 210 µm in diameter, compared with the river sand 68%!.

Profile description

Ap	0-25 cm	Dark grey (10YR4/1.5) loam; strong, medium and fine subangular to angular blocky structure, coarser in the lower part of the plough layer; clear, smooth boundary.
ACg	25-35 cm	Grey (2.5Y5/1) loam; few, faint, fine, yellowish brown mottles (10YR5/4); thick platy structure, caused by sedimentary stratification (loam separated by very thin layers of very fine sand); abrupt, smooth boundary.
Dg	35-120 cm plus	Grey (5Y5/1) uncoated fine sand; in the upper part common, distinct, medium yellowish brown (10YR5/5), mottles, in the lower part no mottles; some cockles (*Cardium edule*), many sand-sized shell fragments; no striking stratification.

Analytical data

Horizon	Depth (cm)	CaCO₃ (%)	pH-KCl	Organic fraction			
				O.m. (%)	C(%)	N(%)	C/N
Ap	0-18	12.5	7.6	2.1	1.54	0.13	11.8
ACg	30-35	15.8	7.7	1.3	n.d.	n.d.	n.d.
Dg	45-60	7.7	8.1	0.3	n.d.	n.d.	n.d.

Horizon	Particle-size distribution of mineral fraction (% of fine earth)						K-fix. (%)
	< 2	2-16	16-50	50-105	105-150	> 150 μm	
Ap	22	11	22	18	23	4.5	13
ACg	21	12	28	25	12	1.5	n.d.
Dg	1.5	0.5	3.5	28	53	13.5	n.d.

Detailed particle-size distribution of sand fraction of the Dg sample

50-75	75-105	105-150	150-210	>210 μm
5	23	53	13	0.5

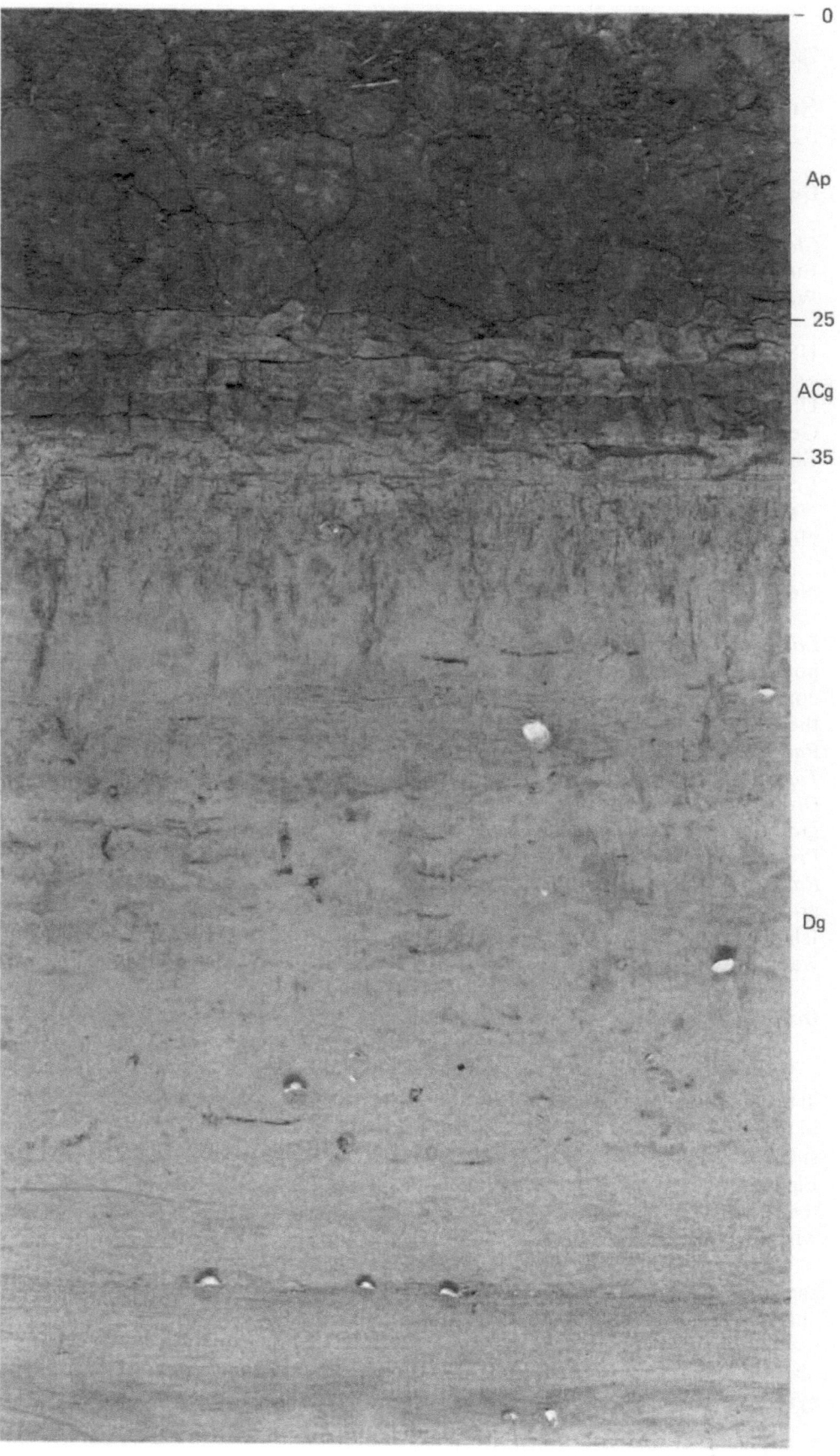

0

Ap

25

ACg

35

Dg

Soils of the marine district: M5

General data

Classification

Eur.	Alluvial soil (p. 4 and 75-77)
World	Calcaric Fluvisol (p. 33) or Calcaric Gleysol (p. 33), coarse textured, level (p. 5)
USA '38	Bog soil, Coastal Marshland (p. 1000 and 1132) or Regosol (p. 120 in the 1949 modification)
USA '75	Siliceous, mesic Typic Psammaquent (p. 185 and 186)
Ger.	Marsch (p. 156-163)
Eng. & Wales	Calcareous sandy gley soil
France	Sol d'apport alluvial hydromorphe (p. 25)
Neth. '50	Young marshy sandy silt soil; mapping unit 14 (p. 111)
Neth. '61	Calcareous, homogeneous, humus-poor and shallowly moderately humose, young sea clay soil, clay-poor sand; mapping unit 6
Neth. '66	'Vlak' vague soil (p. 157, 158 and 194); mapping unit Zn50A

Location. Noordholland Province, a few km south of Den Helder (see Fig. 1) in the polder 'Het Koegras', embanked in 1825, situated between a narrow strip of coastal dunes in the west, the sea wall against the Wadden Sea in the northeast, in the east a younger polder (1846) and much older polders in the south.
Parent material. Recent marine sands.
Topography and elevation. Level, about 0.5 m above sea level.
Drainage and ground water. No tile drains, ditches spaced 100-150 m apart; ground-water level fluctuates between 20 and 70 cm depth.
Present land use. Grassland.
Range in land use. Until very recently mainly grassland, but nowadays bulb growing is becoming more important, except on the same soils on the Frisian Islands, where grassland is maintained because of the lack of fresh water (brackish water in the ditches).

Discussion

Soils with sand at shallow depth, for example the previous soil (M4) are not rare in the polders, but soils completely without a cover of clayey material are very rare in the alluvial part of the Netherlands. They occur as small patches among soils such as soil M4, and mostly are mappable only at scales larger than 1:10 000. On Figure 2 small areas are indicated (mapping unit 3) in the Zuyder Zee Polders, on the Frisian Islands, and south of these islands in an area on the mainland (from which soil M5 was sampled).

This kind of soil can be typified as follows: young (recently embanked from tidal sand flats), marine, calcareous, level, with a controlled ground-water table at shallow depth.

On the Soil Map of Europe these characters justify the name Alluvial Soil. This class may be sandy (Dudal et al., 1966, p. 75), but in the German and French systems nothing is said about sandy textures. According to Mückenhausen (1977,

Table on p. 162) such a soil is called a *Feinsand-Marschboden* in Schleswig-Holstein. In the legend of the Soil Map of the World it might be a Fluvisol according to the key: 'developed from recent alluvial material' (FAO, 1974, p. 44), but the soil satisfies none of the properties listed on p. 32 and the organic matter content is too low. So, in concept it is a Fluvisol, but according to the strict definitions, a Gleysol. In the system of England and Wales it is quite clear that the texture comes first: it is a sandy gley soil. It is clearly a Typic Psammaquent like the soil described in Soil Taxonomy (SSS, 1975, p. 186); the main differences being the soil temperature regime (mesic versus hyperthermic) and the hydrological regime (embanked versus subject to tidal flooding).

Clearly the immediate topsoil has not been ploughed or cultivated for some time and a new turf mat (A11g) has developed in a former plough layer (Apg). The subsurface horizon is little affected by pedogenic processes but is calcareous and mottled as indicated by the designation C2g. The deeper 'bluish' horizon is traditionally called a G horizon. These two horizons have to be coded Cg and Cr (FAO, 1974, p. 22 and 23); C2g and G according to the Soil Survey Manual (SSS, 1951, p. 180 and 181); Go and Gr (Arbeitsgemeinschaft Bodenkunde, 1971, p. 30). In the British system (Hodgson, 1974, p. 71-83) it is not clear where to distinguish between a Bg and a Cg, but below 70 cm depth the code CG has to be used; in the French system (Jamagne, 1967, p. 29 and 36; CPCS, p. 12 and 13) both horizons most probably have to be coded CG, because g is restricted to pseudogley phenomena. In the new USA system a Cg horizon must show 'intense reduction of iron' and 'a base chroma of 1.0 or less, with or without mottles, which is indicative of strong gleying' (SSS, 1975, p. 461), but, on the other hand it is stated on the same page: 'horizons of low chroma in which the color is due to uncoated sand and silt particles are not considered strongly gleyed'. So, either a Cg-Cg or a C-Cg horizon sequence is present.

On the 4th of April 1964, the ground-water level was 45 cm below the surface and it was necessary to lower this level by a small well-point dewatering system (Van der Voort & Kraanen, 1971) in order to describe and sample the soil.

The difference in organic matter content between the turf mat (10.7%) and the former plough layer (2.0%) must be ascribed to soil formation under grassland after the use as arable land. The time lapse is unknown; perhaps a decade.

Profile description

A11g	0-10 cm	New turf developed in former plough layer; very dark grey (10YR3/1) fine sand; few to common, faint, medium yellowish brown (10YR5/5) mottles; single grain to weak very fine subangular blocky structure; clear, smooth boundary.
Apg	10-30 cm	Former plough layer; grey (10YR5/1) fine sand; few faint, medium brown to strong brown (7.5YR5/5) mottles; single grain to very weak structure; abrupt, irregular boundary (spade marks?).
C2g	30-70 cm	Light grey (2.5Y6/1) uncoated fine sand; common, prominent, medium yellowish red (5YR5/5) mottles; single grain, loose; stratified, many fine gravel-sized shell fragments (not analyzed in the fine-earth fraction!); stratified; very abrupt, very smooth boundary.
G	70-120 cm plus	Grey (10Y4.5/1) fine sand; no brown mottles; the upper boundary is most probably a temporary standstill during sedimentation, characterized by a thin darker somewhat siltier layer and by streaks made by burrowing lugworms (*Arenicola marina*); single grain, loose; stratified, many shell fragments. At about 1.10 m depth there is a layer rich in shells: cockles (*Cardium edule*), periwinkles (*Littorina* sp.), *Hydrobia* sp., *Spisula subtruncata, Scrobicularia plana.*

Analytical data

Horizon	Depth (cm)	CaCO₃ (%)	pH-KCl	Organic fraction			
				O.m. (%)	C(%)	N(%)	C/N
A11g	0-6	0.6	7.0	10.7	4.3	0.43	10.0
Apg	12-24	0.6	7.1	2.0	1.0	0.08	12.5
C2g	35-45	1.8	8.4	0.2	n.d.	n.d.	n.d.
G	80-90	2.1	8.4	0.3	n.d.	n.d.	n.d.

Horizon	Particle-size distribution of mineral fraction (% of fine earth)					
	< 2	2-16	16-50	50-105	105-150	> 150 μm
A11g	3	1.5	2.5	2.5	16	75
Apg	4.5	1	1	3	25	66
C2g	3	1	1	2	31	62
G	1	0.5	1	2.5	34	61

Detailed particle-size distribution of the > 150 μm fraction of the G sample

150-210	210-300	300-420	> 420 μm
46	13	2	0.2

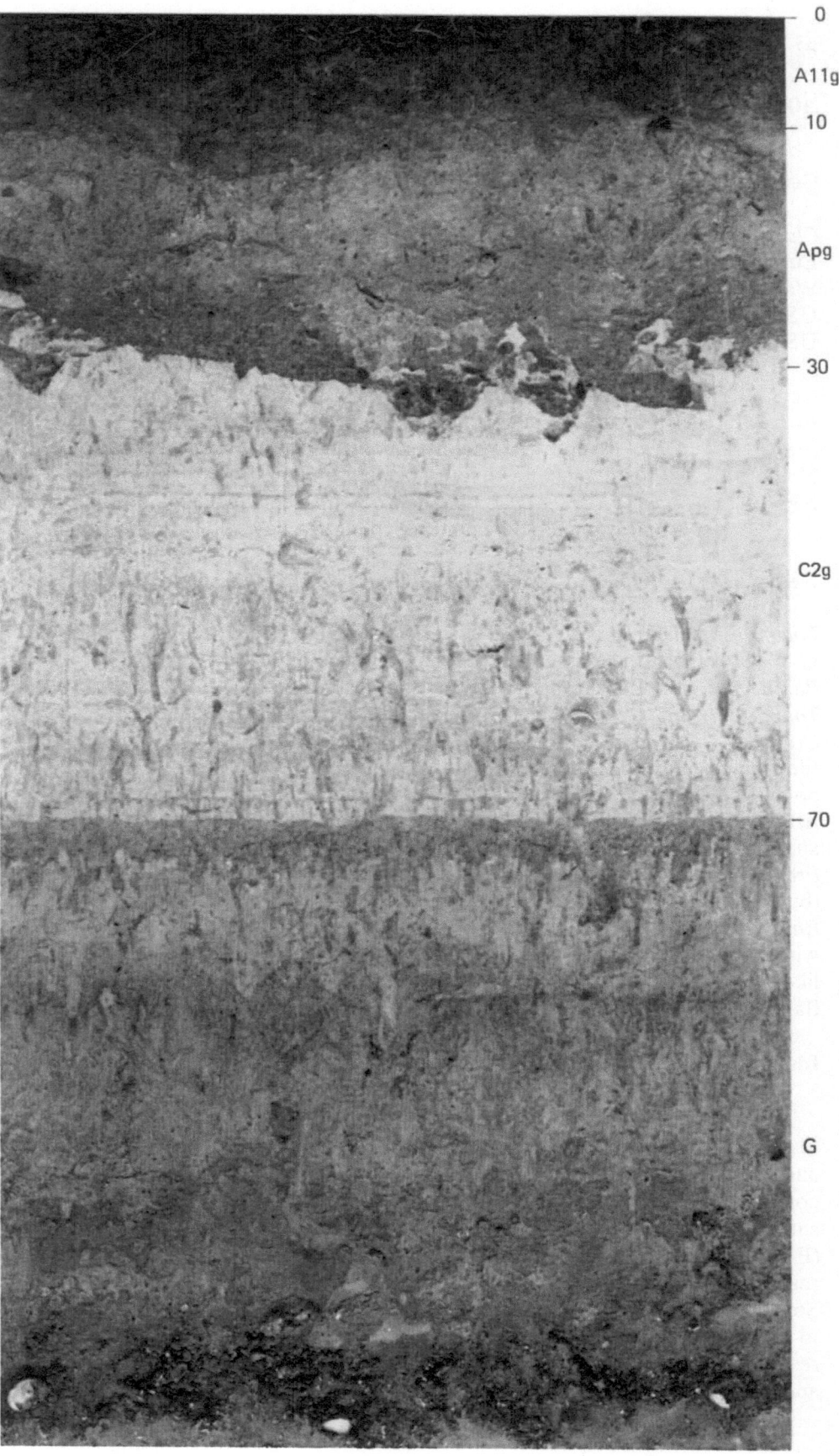

Soils of the marine district: M6

General data

Classification

Eur.	Alluvial soil (p. 4, 75-77)
World	Calcaric Fluvisol (p. 33), medium textured, level (p. 5)
USA '38	Bog soil, Coastal marshland (p. 1000 and 1132)
USA '75	Fine-silty, mixed (calcareous), mesic Typic Fluvaquent (p. 182 and 183)
Ger.	Kalkhaltige Seemarsch (p. 158)
Eng. & Wales	Calcareous alluvial gley soil
France	Sol d'apport alluvial hydromorphe (p. 25)
Neth. '50	Young marsh silt soil; mapping unit 13 (p. 109 and 110)
Neth. '61	Calcareous, becoming lighter with depth, humus-poor and shallowly moderately humose, young sea clay soil, light clay; mapping unit 8
Neth. '66	'Polder' vague soil (p. 160, 161 and 195); mapping unit Mn35A

Location. Zeeland Province, on the isle of Noord-Beveland (see Fig. 1), northeast of Walcheren, in the Wissenkerke Polder, embanked in 1652.
Parent material. Recent marine deposits, calcareous.
Topography and elevation. Level, about 1 m above sea level.
Drainage and ground water. Drained by ditches spaced 80-100 m apart, the field is tile drained about 20 m apart. The polder discharges its water through two younger polders by means of a pumping station into one of the former estuaries of the River Scheldt. The ground-water level fluctuates between 0.5 m in winter and 1.50 m in summer.
Present land use. Used as arable land since embankment but at sampling fallow (building site).
Range in land use. Nearly exclusively arable with dominantly small grains (mainly winter wheat and some malting barley), sugar beet and potatoes, furthermore crops like peas, beans, onions, seed potatoes, caraway, beet seed, grass seed, colza and flax; locally fruitgrowing is important, mainly apples.

Discussion

This is a very typical example of soils in the marine polders. It is medium-textured with a texture profile grading from silt loam to loam, calcareous, ripened and oxidised deeper than 1.20 m (in fact to a depth of 1.50 m), organic matter content is relatively low in the plough layer (2.7%) but in the subsoil there is still sufficient (0.6-0.7%) in this 300-years old polder to satisfy the fluvic characteristics (FAO, 1974, p. 32; SSS, 1975, p. 182 left, item 4) for Fluvisols and Fluvaquents respectively. In terms of the legend of the Soil Map of Europe it is a typical Alluvial Soil. Previous discussion in this book (p. 62) indicates that it is not quite clear how to classify this soil in the old USA system. It is not an Alluvial Soil, for those are restricted to riverine areas, and the description of the Coastal Marshes does not apply to this soil (USDA, 1938, p. 1128 and 1132). Of the European systems only the German discriminates at a high level between fluviatile and marine parent material, namely *Auenboden* and *Marschböden*, and this soil is a *Kalkhaltige Marschboden* (Mückenhausen et al., 1977, p. 158). In the system of England and Wales 'the origin of the soil material (e.g. marine or river alluvium, . . .' (Avery, 1973, p. 335, item 3) is a criterion for differentiating soil series. According to this system it is a calcareous alluvial gley soil, not a typical one, for that subgroup has to be non-calcareous whereas in the Netherlands the calcareous soil is considered typical.

In the French system it is a *sol peu évolué non climatique* at subclass level; group *sol d'apport alluvial* with the addition *hydromorphe* at subgroup level. At the next lower level, the family, 'one can subdivide the soils of a subgroup taking into account the parent material' (CPCS, 1967, p. 9).

The systems of soil horizon nomenclature considered in this book all agree about the use of the letter suffix p for a cultivated topsoil. The lower horizons may yield more discussion about the presence of a Bg or a Cg horizon (see also the discussions on p. 67, 71 and 91). In all these cases the structures are very weak, so most probably these will be C horizons in most systems, although there is a little stratification remaining in the subsoil. The problem of stratification and structure in alluvial soils is discussed at length by De Bakker (1971). In Soil Taxonomy (SSS, 1975, p. 34) horizons with both hydromorphic and fluvic characteristics are excluded from the definition of a cambic horizon. It is to be expected that such horizons are excluded also from the concept of the B horizon, because of item 4 in the definition given in Soil Taxonomy (SSS, 1975, p. 460).

The Dutch system of soil horizon nomenclature uses the suffix g to indicate rusty mottles in a horizon (De Bakker & Schelling, 1966, p. 64). The FAO system is somewhat more specific: 'Mottling reflecting variations in oxidation and reduction' (FAO, 1974, p. 22) and in the new USA system 'The suffix g is used with a horizon designation to indicate intense reduction of iron. . .' (SSS, 1975, p. 461). Obviously the old concepts of g and G (SSS, 1951, p. 180 and 181) have been amalgamated in the new USA system but remain separate in all other systems.

This soil is completely ripened, not only physically, but also chemically. It has no slushy subsoil as occurs in soil M2, and the amount of adsorbed sodium and magnesium cations are both low compared with the soil from the salting (soil M1, p. 62-64). The Ca-Mg ratio is high, which is normal for calcareous marine soils (compare soil M3, p. 70-72). The potassium fixation of soils derived from marine sediments is always much lower than from those of fluviatile parent materials (compare soils F1 and F3).

Profile description

Ap1	0-22 cm	Dark grey (2.5Y4/2) silt loam; moderate fine subangular blocky structure; friable, slightly hard; abrupt, smooth boundary.
Ap2	22-25 cm	Similar material to horizon above; moderate to strong, medium subangular blocky structure; firm, slightly hard; formerly been ploughed, plough-sole; abrupt, smooth boundary.
C21g	25-50 cm	Grey (2.5Y5/1) silt loam; very weak, very coarse compound prisms, breaking to weak fine subangular peds, grading into a structure without peds, characterized by many pores, called spongy structure; friable, soft to slightly hard; few to common, faint, medium brown mottles; fine shell fragments; gradual, smooth boundary.
C22g	50-80 cm	Transitional horizon; coarser textured; mottling more conspicuous; spongy structure; shells (*Spisula subtruncata*) and fine shell fragments.
C23g	80-120 cm plus	Grey (5Y5/1) loam, stratified with thin sandy layers; spongy structure; slightly plastic, very friable, soft; less mottles; less shell fragments.

Analytical data

Horizon	Depth (cm)	Organic fraction				Particle-size distribution of mineral fraction (% of fine earth)					
		O.m. (%)	C(%)	N(%)	C/N	< 2	2-16	16-50	50-105	105-150	> 150 μm
Ap1	0-22	2.7	1.87	0.16	11.7	26	17	35	18	2.5	1.5
C21g	30-50	1.0	n.d.	n.d.	n.d.	24	14	38	20	4.5	0.4
C22g	50-80	0.7	n.d.	n.d.	n.d.	19	9	34	31	6	0.3
C23g	80-110	0.6	n.d.	n.d.	n.d.	19	7	24.5	35	14	0.8

Horizon	CaCO₃ (%)	pH-KCl	Extractable cations (% of sum)					C.E.C. (meq./ 100 g)	Ca/Mg	K-fix. (%)
			Na	K	Mg	Ca	H			
Ap1	8.8	7.2	n.d.	n.d.	n.d.	n.d.	n.d.	n.d.	n.d.	n.d.
C21g	12.2	7.5	3.3	1.7	3.9	87.8	3.3	18.1	22.5	n.d.
C22g	11.5	7.5	n.d.	n.d.	n.d.	n.d.	n.d.	n.d.	n.d.	n.d.
C23g	10.2	7.8	3.9	0.7	5.4	87.7	2.3	13.0	16.3	11

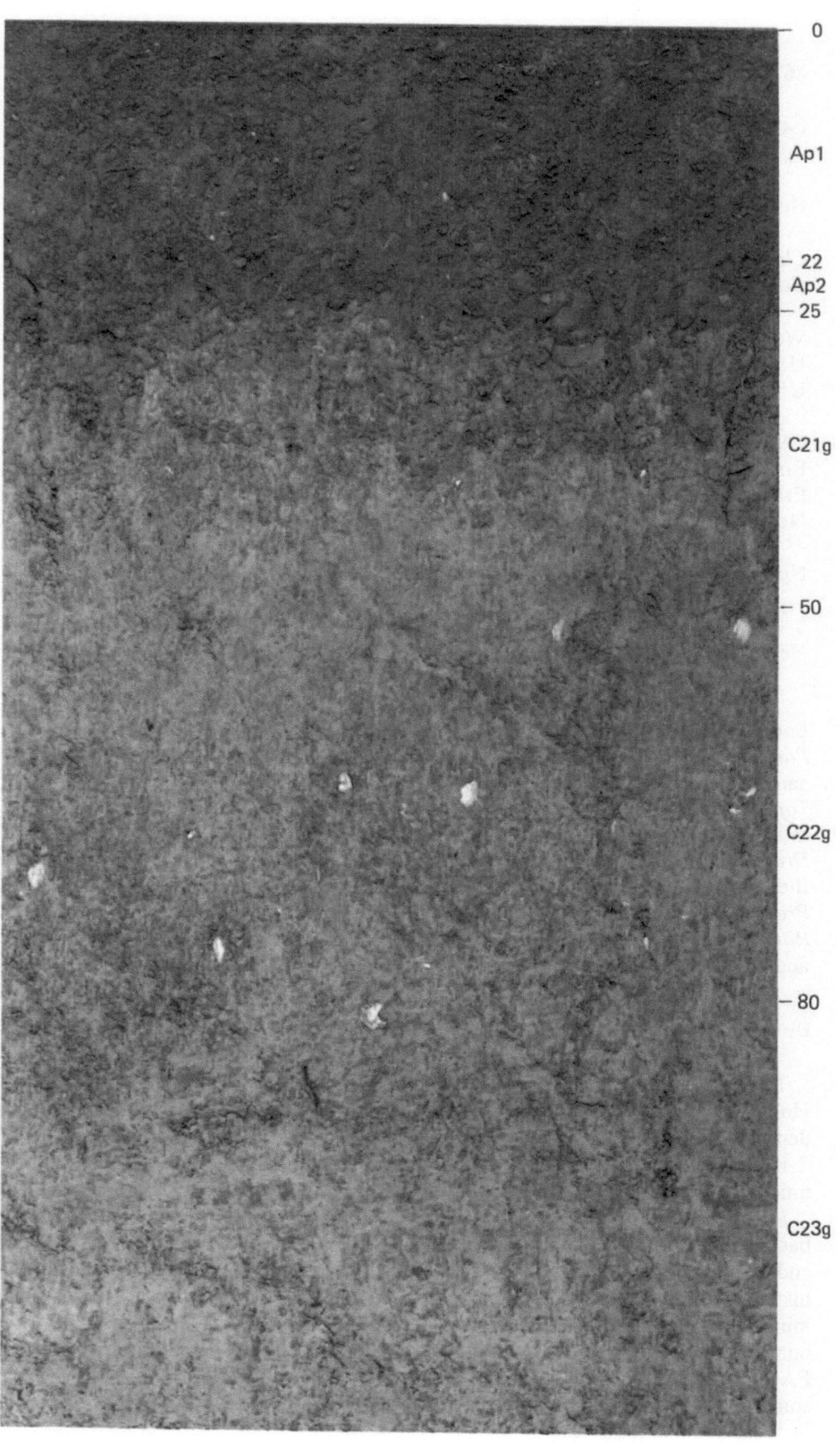

0

Ap1

— 22
Ap2
— 25

C21g

— 50

C22g

— 80

C23g

Soils of the marine district: M7

General data

Clasification

Eur.	No suitable category in the list of soil units (p. 3-7), but on the map included in the association of organic soils (p. 77 and 78)
World	Not possible to classify
USA '38	Not possible to classify
USA '75	Histic over sandy, mesic Haplaquod (p. 336 and 337), proposed to be called both Fluventic and Histic
Ger.	Not possible to classify
Eng. & Wales	Not possible to classify
France	Not possible to classify
Neth. '50	Not possible to classify, but included on the map in the peat soils with clay cover; mapping unit 8
Neth. '61	Undisturbed low moor peat soil, clay underlain by peat within 40 cm with Pleistocene sand in the deeper subsoil; mapping unit 86 m
Neth. '66	'Moer' podzol soil with a clay cover (p. 124 and 188); mapping unit kWp

Location. Friesland Province (see Fig. 1).

Parent material. Recent marine sediment overlying oligotrophic peat over cover sand of Weichsel age.

Topography and elevation. On a very low ridge amidst level polder land; about 0.5 m above sea level.

Drainage and ground water. Drained by ditches spaced 30-60 m apart; the level of the ground water fluctuates between 30 and 90 cm.

Present land use. Grassland used as pasture.

Range in land use. Exclusively used as grassland, with alternating use as pasture and meadow (hay and silage grass).

Discussion

This soil is found in the transition zone where both peat and the overlying Holocene marine sediment is thinning out against the underlying Pleistocene deposits (Fig. 9). Such soils, although widely distributed, are not extensive.

In most systems of soil classification there is no definite category for this rather unique 'soil'. It is two contrasting layers of soil material overlying a buried soil.

However, it is possible to discuss some of the properties of this soil against the background of the diagnostic criteria used in Soil Taxonomy. The peat layer, coded D horizon in the profile description, satisfies item 3 of the definition of the histic epipedon: 'A layer of organic material that ... lies beneath a surface layer of mineral materials that is < 40 cm (16 in.) thick' (SSS, 1975, p.17). Such a thin, buried histic layer is not included in the definition of the Histic H Horizon of the FAO system (FAO, 1974, p.24). In Soil Taxonomy the histic epipedon is used in some places to define Histic subgroups, but no Histic Haplaquod exists.

The soil has a spodic horizon, (SSS, 1975, item 2 of the definition, p.32) with 'its upper boundary within 2 m of the surface' (SSS, 1975, item B1, p.92 and p.333) and because of both of the absence of 'coatings of iron oxides on the individual grains of sand' (SSS, 1975, item BA4a, p.333) and the presence of a histic epipedon the soil is an Aquod; on the level of the great groups it is clearly a Haplaquod. The topsoil is formed from a recent marine sediment; this justifies a second adjective for the name of the subgroup: Fluventic. The particle-size modifier at the family level does not consider this topsoil (SSS, 1975, p.385, item B2); the family must be named 'Histic over sandy'.

On the Soil Map of the Netherlands, scale 1 : 200 000 (Stichting voor Bodemkartering, 1961) the term 'low moor' is used for this soil; in fact, this oligotrophic Sphagnum peat is a drowned high moor.

In our system of soil horizon nomenclature we have not yet introduced Roman numerals for lithologic discontinuities. According to the revised USA system (SSS, 1975, p.460-462) the horizon sequence has to be: A1-AC-IIC-IIIA1b-IIIA2b-IIIB2hb-IIIB3hb. The FAO and British systems differ from the revised USA system in not using Roman numerals. Instead they use arabic figures as prefixes to indicate lithologic discontinuities; the latter system uses the b as prefix to indicate buried soils, whereas the German system uses the f (= fossil) also as a prefix for such situations.

Like all hydromorphic podzol soils in the Netherlands, this buried Haplaquod has a very low iron content (compare soil P1, p.157). The same is true for soils of the Crannymoor Series developed from Shirdley Hill Sand in the Lancashire Coastal Plain in England (Ragg & Clayden, 1973, p.114-117).

The topsoil has an anomalous content in the > 105 micron separate for a sample with this amount of clay (compare soil M3). This is caused by the shallow depth to the cover sand; when the surrounding ditches were dug into this subsoil the spoil was spread over the soil surface.

The high organic matter content of the surface horizon is partly caused by the hydromorphic soil forming conditions on this site, and partly the result of dredging peat from the ditches which increases the amount of sand as well as the organic matter content.

Profile description

A1g   0-8 cm   Very dark grey (10YR3/1) peaty clay; few, faint, fine strong brown mottles (7.5YR5/6); moderate, fine angular blocky structure; clear, smooth boundary.

ACg   8-20 cm   Dark grey (2.5Y4/1) peaty clay; very few, faint, fine dark yellowish brown mottles (10YR4/5); weak medium prismatic structure breaking to fine moderate angular blocky structure; abrupt, smooth to wavy boundary.

D   20-35 cm   Dark reddish brown (5YR2/2) peat; very thin platy; remnants of *Sphagnum spec., Calluna vulgaris* and *Eriophorum;* abrupt, smooth boundary.

A1b   35-45 cm   Black (5YR2/1), peaty fine sand; slightly sticky, slightly plastic to plastic; topsoil of buried podzol developed in Pleistocene sand; gradual, wavy boundary.

A2b   45-60 cm   Grey (7.5YR5/1) fine sand, stained patchy with very dark grey (0YR3/1) vertical streaks of fossil root channels; single grain, loose; gradual, wavy boundary.

B2b   60-75 cm   Dark brown (7.5YR3/3) fine sand with fossil root channels; single grain, very friable; sand grains coated with humus; diffuse, smooth boundary.

B3b   75-120 cm plus   Pale brown (10YR6/3) fine sand with less fossil root channels; single grain, loose; humus coatings becoming thinner with depth, but the C-horizon in this kind of podzol in which grains are not coated (see soil P1) is mostly very deep (1.5-2 m).

Analytical data

Hori-zon	Depth (cm)	Fe$_2$O$_3$ (%)	pH-KC1	Organic fraction				Particle-size distribution of mineral fraction (% of fine earth)					
				O.m. (%)	C(%)	N(%)	C/N	< 2	2-16	16-50	50-105	105-150	> 150 µm
A1g	0-8	n.d.	5.0	23.7	12.3	0.92	13.4	41	15	15	6	10	14
ACg	8-16	n.d.	4.7	16.2	8.3	0.49	16.9	55	21	11	5	3.5	4.5
D	23-33	n.d.	4.4	86.4	n.d.	n.d.	n.d.	n.d.	n.d.	n.d.	n.d.	n.d.	n.d.
A2b	47-56	0.01	4.6	0.6	n.d.	n.d.	n.d.	2	2	0.2	21	33	42
B2b	60-70	0.04	4.6	0.9	n.d.	n.d.	n.d.	0.1	0.7	4.4	21	29	45

Detailed particle-size distribution of the sand fraction of the B2b sample:

50-75	75-105	105-150	150-210	210-300	> 300 µm
9	12	29	30	12	2

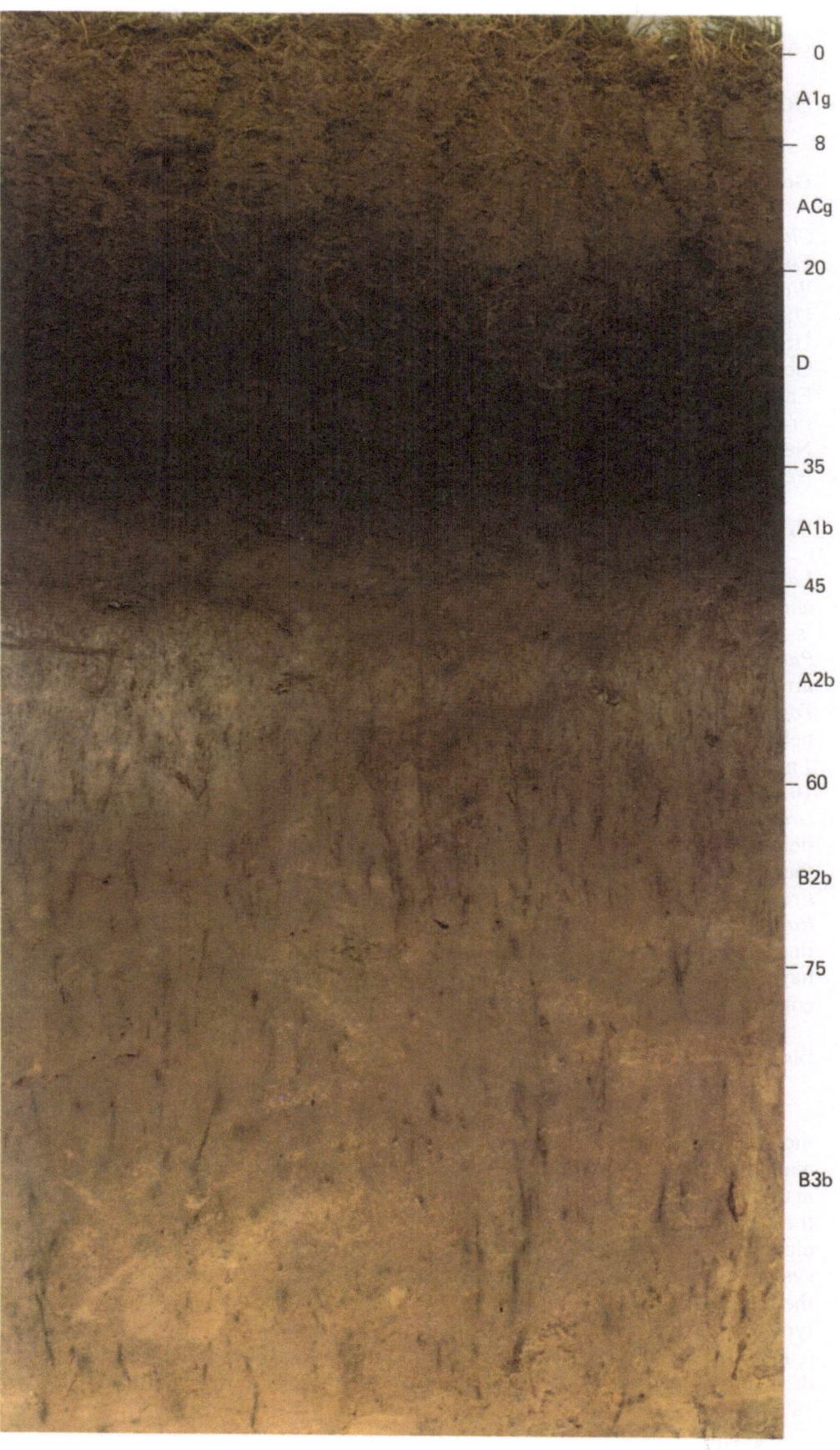

0	A1g
8	
	ACg
20	
	D
35	
	A1b
45	
	A2b
60	
	B2b
75	
	B3b

Soils of the fluviatile district: F1

General data

Classification

Eur.	Alluvial soil (p. 4 and 75-77)
World	Eutric Fluvisol (p. 32), fine textured, level (p. 5)
USA '38	Alluvial soil (p. 1001 and 1133)
USA '75	Fine-clayey, illitic (nonacid), mesic Typic Fluvaquent (p. 182)
Ger.	Pelosol-Auenboden or Pelosol-Vega (p. 148 and 147)
Eng. & Wales	Pelo-(vertic) alluvial gley soil
France	Sol d'apport alluvial hydromorphe (p. 25)
Neth. '50	River basin clay soil; mapping unit 23 (p. 48-52)
Neth. '61	Non-calcareous younger river clay soil, low, homogeneous heavy clay; mapping unit 69
Neth. '66	'Polder' vague soil (p. 160, 161 and 195); mapping unit Rn44C

Location. Gelderland Province, a few hundred meters northwest of the round-about which forms the intersection of the highways Rotterdam-Tiel and Utrecht - 's-Hertogenbosch (see Fig. 1).
Parent material. Recent sediment from the Rhine, protected from flooding by artificial levees or dikes.
Topography and elevation. Level; about 1 m above sea level, in a large backswamp behind the natural levee of the main lower branch of the Rhine (called Waal), about 1 m lower than the natural levee. Approximately 80 km upstream from the sea, and 10 km upstream from the present tidal limit.
Drainage and ground water. Drained by ditches spaced about 80 to 100 m apart, the fields have clear systems of ridges and furrows; the ridges being 10 to 15 m wide; the ground-water level fluctuates between 30 and 100 cm.
Present land use. Grassland.
Range in land use. Almost exclusively used as permanent grassland, with alternating use as pasture and meadow (both hay and silage grass); until about 25 years ago hay meadow predominated. Small areas are forested with poplar and willow coppice (osier); locally duck decoys are present. Some orchards (apple and pear).

Discussion

This soil is a typical example of what has been called formerly an Alluvial Soil (in the old U.S. system and on the Soil Map of Europe). The fact that the parent material has been deposited recently by river water is decisive for its classification in these systems. The same is true for the German and French systems, where also the geogenesis is critical for the classification of this soil. All soils on a recent flood plain are *Auenböden* (also called *Vegas*) in the German system, and in the French system *sols d'apport alluvial* either *sol minéraux bruts* or *peu évolués*. Because of the fine texture (roughly 60% clay) the Germans add *Pelosol* to the name of the type. The presence of 'more than traces of organic matter in the upper 20 cm' (CPCS, 1967, p.21) justifies its place in the class of the *sols peu évolués* and because the depth of the ground water is 'during certain periods at less than 1 m' (CPCS,

1967, p.25), this soil is called *hydromorphe* at subgroup level. However, this name might be wrong, if the subsurface horizon has to be called a (B)-horizon according to the French nomenclature (see below).

In the system of England and Wales this soil is called a ground-water gley soil at the level of major group (the highest level) and at a group level an alluvial gley soil, based on the fact that its parent material is recent alluvium, and like the Germans, the English add pelo-(vertic) at subgroup level.

The absence of progressive soil formation and presence of the two buried A1-horizons with the organic matter content over 0.35% is diagnostic of the 'fluvic' characteristics; Fluvisol in the legend to the Soil Map of the World and Fluvaquent in Soil Taxonomy. The adjective Eutric in the former system results from the high base saturation, a diagnostic criterion lacking in Soil Taxonomy. On the other hand this system subdivides the fluvic soils into Fluvaquents and Fluvents, an essential differentiation, for not all Fluvisols are gleyic. The other additions in the names need no further discussion.

The subsurface horizons show hydromorphic properties but do not show fine sedimentary stratification like soil LP 3, and have a clearly developed soil structure. It is not always clear if such horizons should be called Bg or Cg in some international systems of soil horizon nomenclature. Only in the German system is there no such difficulty as the subsurface horizons have to be called Go-horizons (compare soil 49 in Mückenhausen et al., 1977, p.274).

The indication Bg or Cg has no relevance in the classification of the USA, English and FAO systems, but it does in the French. The *sols peu évolués* are not allowed to have a B-horizon. 'In these soils there are never A2 horizons, B horizons, even no (B)horizons' (CPCS, 1967, p.21). If this definition is used consistently all alluvial soils showing structure in a gleyed subsurface horizon must be classified as *sols à gley*, and the correlation as *sol d'apport alluvial hydromorphe* is incorrect.

The analytical data are characteristic of fine-textured soils from the backswamp areas of the lower Rhine and Meuse flood plain. In contrast to the fine-textured soils developed from marine sediments (e.g. soil M3 and LP2), these soils have a high potassium fixation capacity. This property is related to the relatively large amount of vermiculite in this soil compared with the traces found in soils developed from marine sediments, and probably also to a special kind of smectite which has a higher potassium fixation than smectite in marine sediments.

Profile description

A1g	0-8 cm	Dark grey brown (10YR3.5/2) silty clay; few mottles; moderate coarse compound prisms, breaking to moderate very fine subangular blocky structure; friable, slightly hard; abrupt, smooth boundary.
ACg	8-25 cm	Transitional horizon with less and coarser blocks in the compound prisms.
C1g	25-50 cm	Dark grey (10YR4/1) clay; few, faint, medium mottles; strong, very coarse prisms, breaking to coarse, blocky peds; firm, slightly hard to hard; abrupt, smooth boundary.
A11b	50-64 cm	First buried topsoil (Roman Age?); dark grey (N4) clay; non-calcareous but some secondary lime in fossil root channels.
C11bg	64-94 cm	Grey (5Y4.5/1) silty clay; common, faint to distinct, fine, yellowish brown (10YR5/4) mottles; moderate, very coarse prismatic structure; firm, hard; abrupt, smooth boundary.
A12b	94-106 cm	Second buried topsoil (Bronze Age?); dark grey (N4) clay to silty clay.
C12bg	106-120 cm plus	Grey (5Y5/1) silty clay; few mottles; weak very coarse prisms, below about 1,10 m apedal, i.e. cracks absent; soft and not completely ripened.

Analytical data

Horizon	Depth (cm)	Organic fraction				Particle-size distribution of mineral fraction (% of fine earth)			
		O.m. (%)	C(%)	N(%)	C/N	< 2	2-16	16-50	> 50 μm
A1g	0-5	16.8	8.5	0.77	11.0	52	18	23	6.5
ACg	8-20	6.5	3.4	0.36	9.4	59	17	19	5.5
C1g	30-40	2.4	n.d.	n.d.	n.d.	68	24	7	0.8
A12b	96-106	1.9	n.d.	n.d.	n.d.	54	23	17	6.0

Horizon	CaCO₃ (%)	pH-KCl	Extractable cations (% of sum)					C.E.C. (meq./ 100 g)	Ca/Mg	K-fix. (%)
			Na	K	Mg	Ca	H			
A1g	0.0	4.8	n.d.	n.d.	n.d.	n.d.	n.d.	n.d.	n.d.	n.d.
ACg	0.0	4.6	n.d.	n.d.	n.d.	n.d.	n.d.	n.d.	n.d.	n.d.
C1g	0.2	5.9	0.9	1.1	9.5	82.8	5.6	46.5	8.6	88
A12b	0.3	6.1	n.d.	n.d.	n.d.	n.d.	n.d.	n.d.	n.d.	58

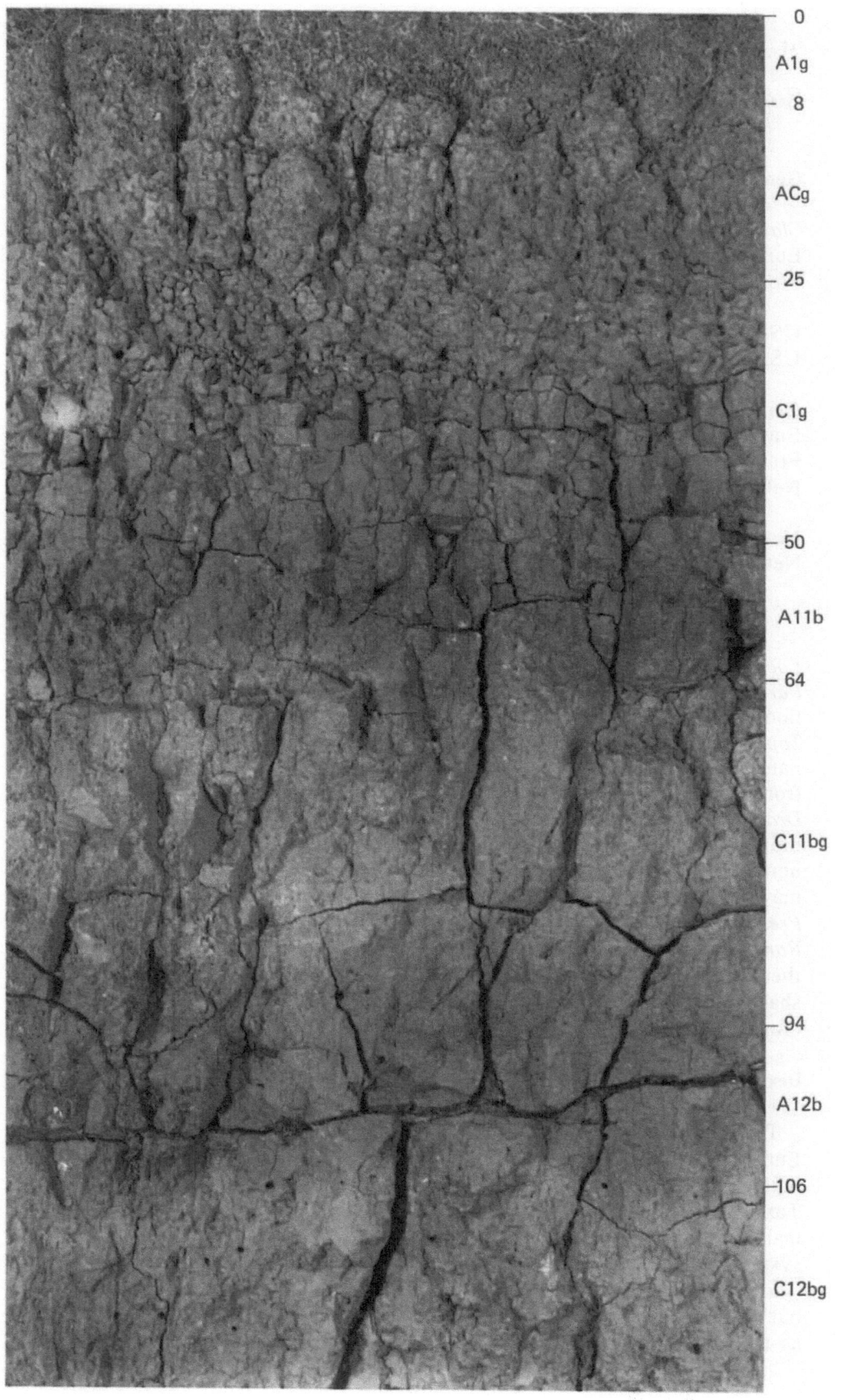

0

A1g

8

ACg

25

C1g

50

A11b

64

C11bg

94

A12b

106

C12bg

Soils of the fluviatile district: F2

General data

Classification

Eur.	Alluvial soil (p. 4 and 75-77)
World	Calcaric Fluvisol (p. 33), medium textured over coarse textured, level (p. 5)
USA '38	Alluvial soil (p. 1001 and 1133)
USA '75	Fine-loamy over sandy, mixed (calcareous) mesic Fluvaquent (p. 182-184), proposed to be called Psammic
Ger.	Auenboden (p. 141-149)
Eng. & Wales	Calcareous sandy gley soil
France	Sol d'apport alluvial hydromorphe (p. 25)
Neth. '50	Included in the calcareous river levée soils; mapping unit 25, mentioned on p. 41 as Rs1
Neth. '61	Not mapped separately, is an inclusion in mapping unit 66
Neth. '66	'Vlak'vague soil (p. 157, 158 and 194), not mapped separately in the fluviatile area because they are minor inclusions in mapping unit Rd90A

Location. Gelderland Province, about 30 km southeast of Utrecht (see Fig.1).

Parent material. Recent Rhine sediment: shallow loam over sand; protected from flooding by artificial levees.

Topography and elevation. Level; about 4 m above sea level. Situated on the natural levee of one of the lower branches of the Rhine, about 100 km upstream from the North Sea and about 30 km upstream from the tidal limit.

Drainage and ground water. Not tile drained but drained by ditches into main drains and by pumping station into the river. The ground-water level fluctuates normally between 1 and 2 m depth, but when the river is abnormally high, there may be seepage underneath the dike through the coarse subsoil (Fig. 35).

Present land use. Grassland, arable until 5 years ago.

Range in land use. The soils only occupy patches amidst deeper soils (like soil F3), then the range in land use is the same as of soil F3. If larger parts of the fields have shallow sand, early potatoes and spring grains (oats and barley) are the dominant crops.

Discussion

This soil is also an Alluvial Soil in the old USA system and on the Soil Map of Europe using the same arguments as in the previous discussion. The shallow depth of the sand causes some difficulties in other systems of classification. Both in Soil Taxonomy and in the FAO-legend the criterion for fluvic characteristics (organic matter content more than 0.35%) is not valid for sandy layers. Only the latter system has a criterion which justifies its classification as a Fluvisol, namely 'showing fine stratification' (second part of the definition of the Fluvisols on p. 32), but so have most Regosols! This criterion is lacking in the recent USA system but as the loamy topsoil is too thick for a Psammaquent, a Fluvaquent must be advocated. In both systems the criteria fail to give a solution and the old argument (recent alluvium) is decisive again. It is a Calcaric Fluvisol (World Soil Map) based on the lime content, and must be placed in a Psammic subgroup in Soil Taxonomy based on the presence of sand at shallow depth.

In the system of England and Wales the alluvial gley soils are texturally restricted to 'loamy and clayey alluvium', but the definition of the concept 'sandy soils' (Avery, 1973, p. 334) classifies this fluviatile soil at subgroup level as a calcareous sandy gley soil.

In the French system there might be the same discussion about this soil as about the two other soils from the fluviatile district: questioning designation the AD horizon as a (B) horizon?

In the German system there is also no reference to shallow soils and it is difficult to know in which kind of *Auenboden* to place this soil.

The coarse subsoil is coded Dg horizon, in most systems this will be a IICg. Sometimes this rapidly permeable material is connected via the subsoil with the actual river bed (Fig. 35).

The meander belts of the Lower Rhine and Meuse are characterized by these sandy subsoils between 80 and 150 cm depth, gravels seldom being present in the first few meters. River sands are considerably coarser than sea sands in the Netherlands (compare soils M4, p. 76 and soil M5, p. 80).

Fig. 35. Wet spots in shallow soils caused by seepage underneath the dike, after an exceptionally high river level in early summer.

Profile description

A11	0-3 cm	New turf developed in former plough layer; very dark greyish brown (10YR3/2) loam; no mottles; moderate, thin platy structure breaking to fine subangular blocky peds; friable, slightly hard; abrupt, smooth boundary.
Ap	3-30 cm	Former plough layer, dark greyish brown (10YR4/2) loam; no mottles; weak, coarse compound prisms, breaking to weak, fine to very fine subangular blocky structure; friable, slightly hard; clear, smooth to irregular boundary.
AD	30-40 cm	Brown (10YR4.5/3), loamy sand, no mottles; clear, wavy boundary.
Dg	40-120 cm plus	Grey (10YR6/1.5) sand; common, faint, coarse, dark brown to strong brown (7.5YR3/2-5/6) mottles; single grain; loose; stratified with coarse layers containing coarse fragments just over 2 mm.

Analytical data

Horizon	Depth (cm)	CaCO₃ (%)	pH-KC1	Organic fraction			
				O.m. (%)	C(%)	N(%)	C/N
Ap	3-30	0.1	6.5	2.0	1.02	0.12	8.5
AD	30-40	7.1	7.9	0.7	n.d.	n.d.	n.d.
Dg	60-75	2.5	8.4	0.2	n.d.	n.d.	n.d.

Horizon	Particle-size distribution of mineral fraction (% of fine earth)					K-fix (%)
	< 2	2-16	16-50	50-105	> 150 µm	
Ap	21	14	20	7	38	33
AD	8	3.5	7	9	73	n.d.
Dg	0	0	2.5	0.5	97.5	n.d.

Detailed particle-size distribution of the > 105 µm fraction of the D sample:

105-150	150-210	210-300	300-420	420-600	600-850	> 850 µm
3.5	26	41	18	4.5	2.5	2

0
A11
3

Ap

30

AD

40

Dg

Soils of the fluviatile district: F3

General data

Classification

Eur.	Alluvial soil (p. 4 and 75-77)
World	Calcaric Fluvisol (p. 33) or Calcic Cambisol (p. 38)
USA '38	Alluvial soil (p. 1101 and 1133)
USA '75	Fine-silty, mixed, mesic Fluventic Eutrochrept (p. 251 and 252)
Ger.	Braunerde-Auenboden (p. 148)
Eng. & Wales	Typical brown calcareous alluvial soil
France	Sol d'apport alluvial modal (p. 25) or sol brun calcaire modal (p. 37)
Neth. '50	Calcareous river levée soil, mapping unit 25, river ridge soil in the book (p. 40-47)
Neth. '61	Calcareous younger river clay soil, high, becoming lighter with depth, moderately sandy clay; mapping unit 66
Neth. '66	'Ooi'vague soil (p. 163 and 195); mapping unit Rd90A

Location. Gelderland Province, 15 km west of Wageningen (see Fig.1).
Parent material. Recent Rhine sediments, protected from flooding by artificial levees or dikes.
Topography and elevation. Level; about 7 m above sea level. Situated on the natural levee of a former Rhine distributary about 120 km upstream from the coast, about 3 km south of the actual (embanked) Rhine course.
Drainage and ground water. Few ditches, no tile drains; ground-water table always below 1.30 m depth.
Present land use. Apple orchard, fallow round the trees, between the rows of the fruit trees the field is under grass, which is 'mulched', i.e. mowed several times a year and left to decay.
Range in land use. Arable, orchard and pasture. The main crops are winter wheat, potatoes and sugar beet; locally there is a wider crop rotation with silage maize, beans, peas and seed grass. In the last half century fruit growing became important, mainly apples and cherries, and to a lesser extent pears and plums; but in the last decade the area devoted to orchards, has decreased, because of the high harvesting costs (particularly cherries). Small areas are used for red and black currents, strawberries, and shrub and tree nurseries. As on many soils in the Netherlands the area of grass (long-term leys) has increased in the last 10 years.

Discussion

This soil is a classic example of the brown, deep loamy, well-drained soils with a favourable structure of the natural levees of the Rhine. Clearly it is an Alluvial Soil in the old USA system and on the Soil Map of Europe. In the first edition of the German system it was an *autochthone braune Kalkvega* or *autochthoner brauner Kalkauenboden* (Mückenhausen et al., 1962, p.127). In the second edition (Mückenhausen et al, 1977, p.148) it is a transition between a *Auenboden* and a *Braunerde: a Braunerde-Auenboden.*

These three systems have the origin of the parent material at the highest level of consideration but the Eng. & Wales system classifies this soil at major group level as a brown soil, considering at group level the parent material. Formerly the English would have spoken of a brown warp soil (Rag & Clayden, 1973, p.48) but now this soil is called a typical brown calcareous alluvial soil.

Although the French system has two classes for soils lacking well developed horizons, *'sols minéraux bruts'* and *'sols peu évolués'*, this soil has to be grouped with the class of the *sols calcimagnésiques* because of the presence of the brown calcareous subsurface horizon, and at subgroup level as a *sol brun calcaire modal*.

Fluvisols do not have a Bw horizon which meets the definition of the Cambic B horizon (FAO, 1974, p.26). The subsurface horizon of this soil is a Cambic B horizon and so this soil is not a Fluvisol but a Cambisol, and because it is calcareous it is a Calcic Cambisol. The same arguments are valid in Soil Taxonomy and the description of the Fluventic Eutrochrept exemplifies exactly this Dutch soil. 'These soils formed in calcareous alluvium in flood plains along streams and rivers. The content of organic carbon ... is relatively high at depth'. (Soil Survey Staff, 1975, p.252).

In most systems of soil horizon nomenclature the C21 horizon must be called a B, Bw, (B) or Bv horizon. Traditionally subsurface horizons without indications of illuviation never have been coded B horizon in the Netherlands. However, there is a sufficient evidence of alteration to satisfy definitions of B horizons in other systems of soil horizon nomenclature. Of these, obliteration of the original stratification is the most conspicious phenomenon in this soil. The process of gradual disappearance of the original sedimentary stratification by burrowing animals such as moles and earthworms is called biological homogenization in the Netherlands (Hoeksema, 1953). In the C22g especially (better called BwCg horizon) this process obviously is dominant but other evidence of alteration include stronger chroma and redder hue than the underlying horizon, presence of soil structure and removal of calcium carbonate.

However, differences in lime content may have been initial, due to decalcification during upward growth of the sediment (Van der Sluijs, 1970), as well as by subsequent leaching.

Although potassium fixation in this medium-textured soil is lower than that of the fine-textured soil F1, it is still considerably higher than in soils of marine origin of the same texture (e.g. soils M4 and M6). The C-N ratio of the plough layer is 8.9, a low value which is quite normal for plough layers of medium-textured soils developed in fluviatile sediments. These values are roughly 2.5 point higher on arable land in the marine polders, a fact for which there is no explanation at present.

Profile description

Ap	0-23 cm	Dark greyish brown (10YR3.5/2) silt loam; moderate, fine subangular blocky structure; friable; abrupt, smooth boundary.
C21	23-64 cm	Brown (10YR4/2.5) silt loam; not stratified; no mottles; weak, fine and very fine subangular blocky to crumb structure, many pores; many worm channels; friable; gradual, wavy boundary.
C22g	64-90 cm	Transitional horizon between the homogeneous overlying horizon and the lower stratified layer. Besides stratified, grey (10YR5.5/1) slightly mottled parts, browner and homogenised material predominates; many worm channels coated with dark material (dark streaks on the plate); gradual, irregular boundary.
C23g	90-120 cm plus	Stratified material: thin layers grey (2.5Y5/1) loam, alternating with lighter coloured loamy sand layers; iron staining is conspicuous on the boundaries between the thin strata; some krotovinas.

Analytical data

Horizon	Depth (cm)	Organic fraction				Particle-size distribution of mineral fraction (% of fine earth)					
		O.m. (%)	C(%)	N(%)	C/N	< 2	2-16	16-50	50-105	105-150	> 150 μm
Ap	0-23	4.7	2.4	0.27	8.9	19	17	43	12	4	5
C21	45-55	1.8	n.d.	n.d.	n.d.	24	19.5	34	14	4.5	4
C23g	100-110	0.6	n.d.	n.d.	n.d.	16	12.5	34	24	7	6.5

Horizon	CaCO₃ (%)	pH-KC1	Extractable cations (% of sum)					meq./100 g			Ca/Mg	K-fix. (%)
			Na	K	Mg	Ca	H	C.E.C.	CaO	SO₄		
Ap	3.4	7.3	n.d.	n.d.	n.d.	n.d.	n.d.	n.d.	n.d.	n.d.	n.d.	n.d.
C21	7.0	7.7	1.5	0.5	5.4	90.8	1.8	20.7	n.d.	n.d.	16.8	46
C23g	13.9	7.8	n.d.	n.d.	n.d.	n.d.	n.d.	n.d.	278	1	n.d.	n.d.

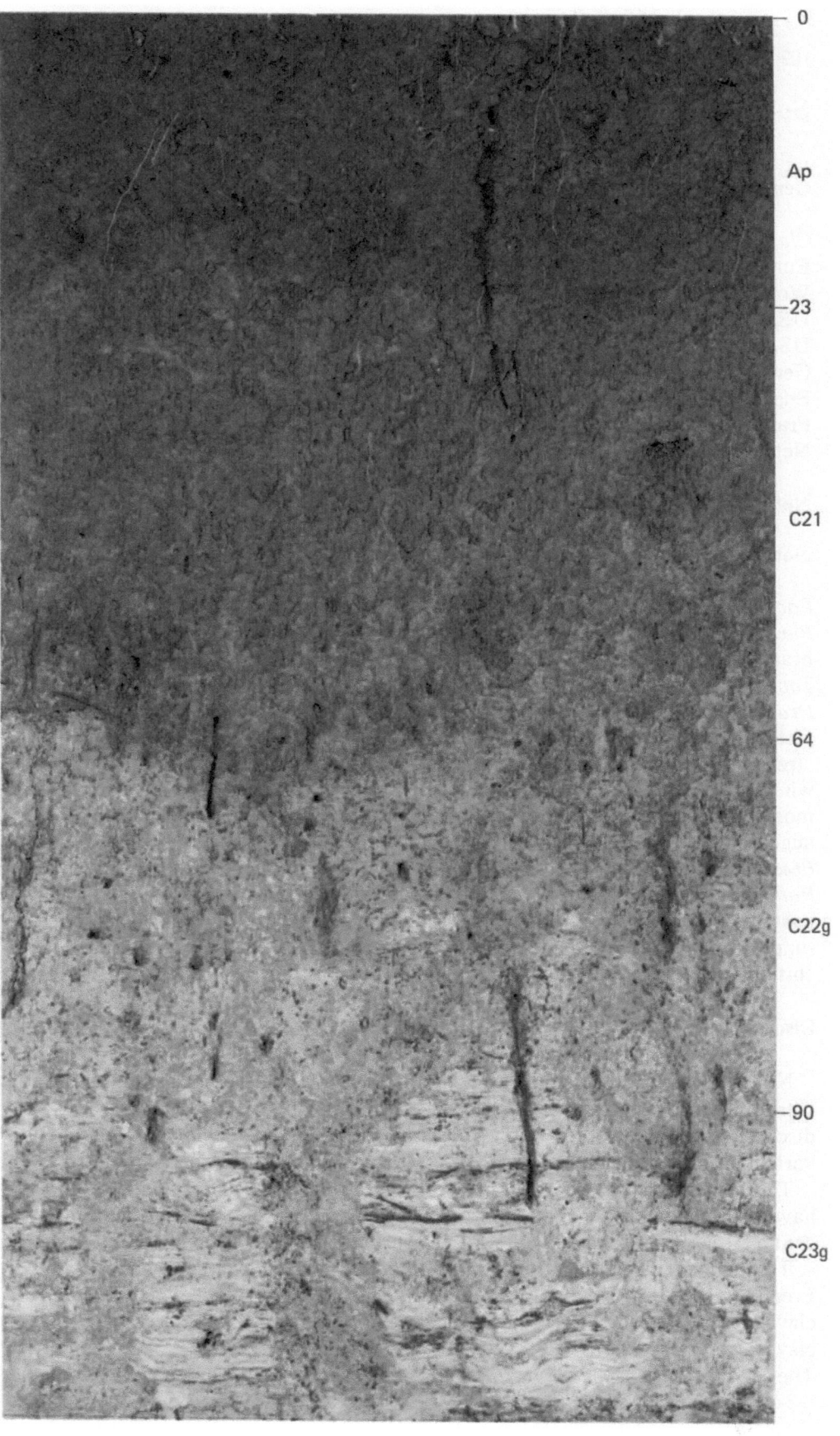

0

Ap

23

C21

64

C22g

90

C23g

Soils of the drained lakes and peat uplands district: LP1

General data

Classification

Eur.	Organic soils (p.7)
World	Eutric Histosol (p.41), level (p.5)
USA '38	Bog Soil, Peat and Muck (p.1000 and 1128-1132)
USA '75	Euic, mesic Typic Medihemist (p.222)
Ger.	Niedermoor (p.167 and 168)
Eng. & Wales	Earthy eutro-amorphous peat soil
France	Sol à tourbe semi-fibreuse, eutrophe (p.74)
Neth. '50	Low wood peat soil; mapping unit 29 (p.65-78, especially 76 and 77)
Neth. '61	Low moor peat, not cut over, overlying woody peat; mapping unit 85
Neth. '66	'Koop' peat soil (p.113 and 185); mapping unit *oh*Vb

Location. Utrecht Province, about 20 km west of the town of Utrecht (see Fig.1).
Parent material. Wood peat, mucky material with big remnants of roots and branches, to a depth of about 6 m, overlying Pleistocene sand.
Topography and elevation. Level; about 1.5 m below sea level.
Drainage and ground water. Drained by wide ditches, spaced 50-60 m apart, the water is discharged into main drainage channels (shipping canals or natural streams) by pumping stations; the ground-water level fluctuates between 30 cm in winter and 70 cm below the surface in summer. During sampling and taking the monolith, which took three hours there was no water in the pit, but in a one-day old auger hole the water level was at a depth of nearly 50 cm.
Present land use. Grassland.
Range in land use. Nearly exclusively used as grassland, with alternating use as pasture and meadow, both hay and silage-grass. Locally flowers in glasshouses (upper right part of Fig.15) like roses, carnations and lilacs, elsewhere ornamental shrubs, e.g. conifers, azaleas.

Discussion

Nearly all the classification systems referred to in this book have a definite category for this soil, a prototype of a low moor peat soil. In the following discussion the soil will be examined in the light of the criteria employed in those various classifications.

There is only one class of organic soils on the Soil Map of Europe, soils must have 'an organic matter content of more than 30 percent to a depth of at least 30 centimetres' (Dudal et al., 1966, p.7). Taken strictly this definition does not fit this soil as the second horizon has less than 30% organic matter. The same is true for the French system in which the required organic matter content depends on the clayeyness: 'more than 30% over at least 40 cm depth if the mineral material is clayey, and more than 20% if the mineral material is sandy' (CPCS, 1967, p.73). The subdivision at the level of the *groupe* is the same as in the new USA system.

Some systems have no clear morphometric criteria, so concepts rather than definitions are decisive for the classification of soils, e.g. Bog soil (USDA, 1938) and *Niedermoor* (Mückenhausen et al., 1977).

In Soil Taxonomy this soil is a Typic Medihemist (SSS, 1975, p.222) and in the FAO system a Histosol. Four systems (French, England and Wales, the new USA and FAO) have two reaction classes, called respectively *mésotrophe et eutrophe*, eutro- (or eu-), euic and eutric versus *oligotrophe*, oligo-, dysic and dystric. The definition of this boundary is either pH 4.5 (0.01 M Ca Cl$_2$) or pH 5.5 (H$_2$O, 1:5), pH values determined in KC1 are used in Holland which give values lower than CaCl$_2$. According to all systems this soil goes into the eutrophic class, but only the system of England and Wales, and the Dutch system (De Bakker & Schelling, 1966, p.174) recognize the earthy topsoil at subgroup level.

Most of the systems of soil horizon nomenclature discussed use special capital and lower case letters for soil horizons of organic soils. The Dutch, however, consider peat as one kind of parent material, like all other kinds of parent material in which, soil horizons 'with characteristics produced by soil-forming processes' occur (SSS, 1975, p.459). Examples of these processes are: ripening, oxidation, weathering, moulding ('earthifying'), influences of man (manuring, adding a sandy topsoil), and humus-illuviation (Van Heuveln & De Bakker, 1972).

In Soil Taxonomy the horizon sequence has to be coded: Oa1-Oa2-Oa3-Oe1-Oe2 (SSS, 1975, p.212); in the FAO system simply H1-H5. In the French system there are no horizon designations for peat soils, neither Jamagne(1967), the CPCS (1967) or Duchaufour(1965) discuss this problem. In the German system (Arbeitsgemeinschaft Bodenkunde, 1971, p.31) these horizons have to be coded Hn1-Hn5 and in the British system: Oh1-Oh2-Oh3-Om1-Om2 (Hodgson, 1974, p.73). Sealé (1975, p.71), when surveying the soils of the Fenlands between Cambridge and The Wash, evolved a different nomenclature from the official system when he defined horizons in soils derived from organic matter. 'Aop-ploughed layer of peaty soil or peat; Bo=humified subsurface horizon, distinguished by its structure; Co = little altered peat substratum'.

The use of the letter subscript an (=anthropogenic) can be explained by the high sand content of the topsoil. This is ascribed to certain management practices in this area. Sand has been used as absorbing substance in the stables, and when mixed with spoil dredged from the ditches this earth-containing manure was spread over the fields. The C-N ratio is low, which is normal for soils developed from wood peat; other data from the same kind of peat show that this ratio increases to 20-25 in the non-oxidised subsoils.

Profile description

Aang	0-13 cm	Very dark brown (10YR2/2) peaty clay loam; many, faint, fine reddish brown (5YR5/6) mottles; compound moderate, very fine subangular blocky structure; very friable when moist. The suffix *an* (anthropic) is justified by the presence of fragments of pottery, bricks and glass, cinder and many sand grains; clear, smooth lower boundary.
A12g	13-30 cm	Very dark grey brown (10YR3/1.5) peaty clay; common, distinct, fine strong brown mottles; compound moderate, medium prisms, breaking to strong, fine, subangular blocky structure; firm when moist; fewer anthropic inclusions, some sand grains.
ACg	30-40 cm	Gradual and smooth transition from the moulded clayey topsoil to the non-moulded and less clayey subsoil.
C1	40-65 cm	Black (5YR2/1) peat, oxidised wood peat with common, prominent, coarse fragments of wood and many other recognisable plant remains (branches, roots, leaves) in a matrix of non-recognisable mucky material; very clear and smooth lower boundary.
G	65-120 cm plus	Dark reddish brown (5YR3/3) peat; changing very rapidly into black (5YR2/1) on exposure to air; non-oxidised, intensely reduced wood peat with more plant remains which are more clearly recognisable than in the C1-horizon.

Analytical data

Horizon	Depth (cm)	Organic fraction				Particle-size distribution of mineral fraction (% of fine earth)			
		O.m. (%)	C(%)	N(%)	C/N	< 2	2-16	16-105	>105 μm
Aang.	0-5	45.3	n.d.	n.d.	n.d.	33	12	30.5	24.5
A12g	15-20	23.1	11.3	1.19	9.6	54	10	24	12
G	80-90	67.4	n.d.	n.d.	n.d.	46	15	36	3

Horizon	CaCO₃ (%)	pH-KC1	Extractable cations (% of sum)					C.E.C. (meq./ 100 g)
			Na	K	Mg	Ca	H	
Aang	0.0	5.2	n.d.	n.d.	n.d.	n.d.	n.d.	n.d.
A12g	0.0	5.2	1.5	1.6	6.0	67.6	23.3	61.5
G	0.1	5.8	n.d.	n.d.	n.d.	n.d.	n.d.	n.d.

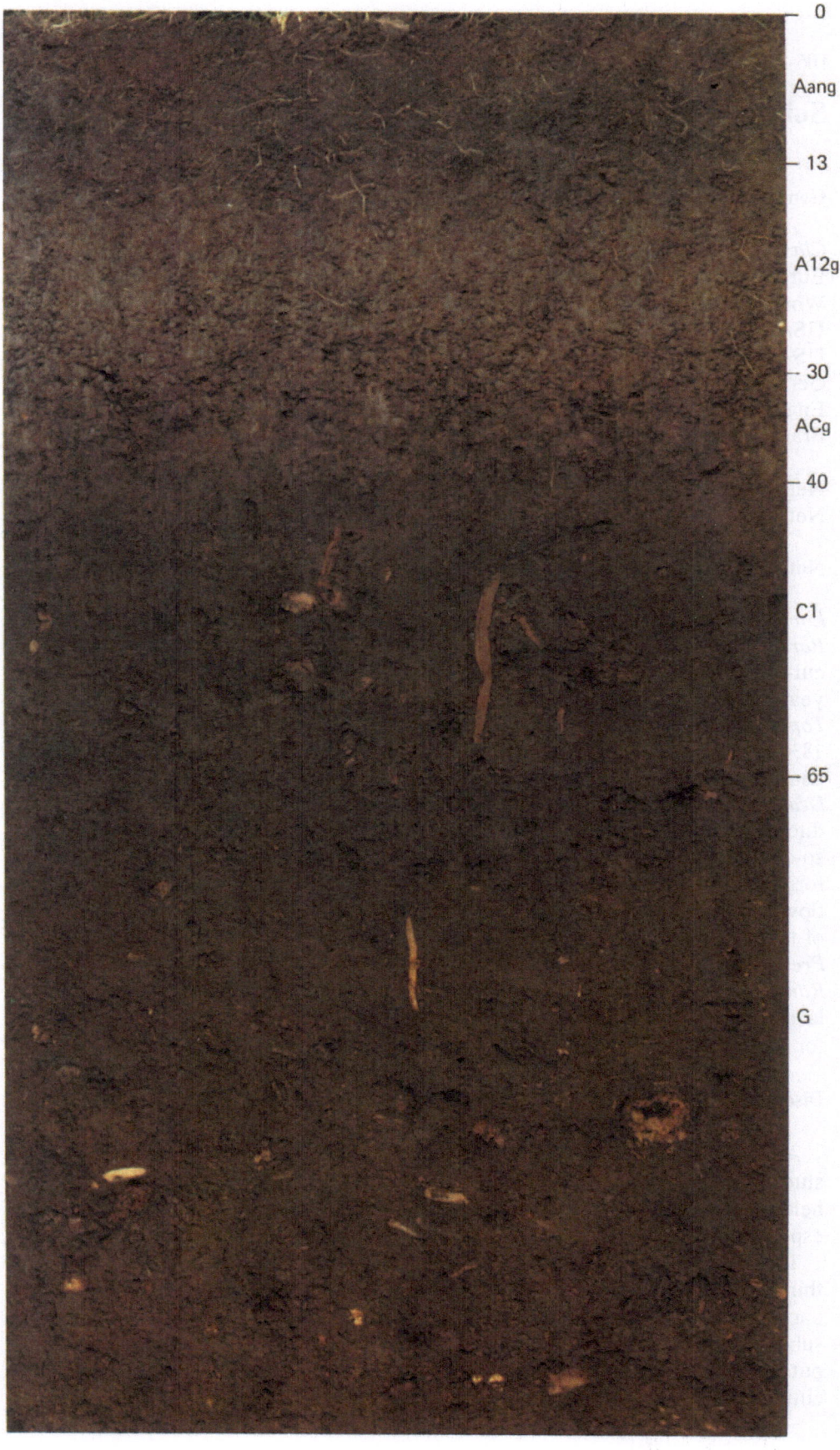

0

Aang

— 13

A12g

— 30

ACg

— 40

C1

— 65

G

Soils of the drained lakes and peat uplands district: LP2

General data

Classification.

Eur.	Included in the Alluvial soils (p.4)
World	Thionic Fluvisol (p.33), fine textured, level (p.5)
USA '38	Half-Bog soil (p.1000 and 1110)
USA '75	Histic over clayey, mesic Sulfaquept (p.245)
Ger.	Organomarsch (?) (p.162 and 163)
Eng. & Wales	Sulphuric humic alluvial gley soil
France	Sol humique à gley (p.75), not possible to subdivide at subgroup level
Neth. '50	Non-calcareous soap clay with 'katteklei'; mapping unit 4k
Neth. '61	Non-calcareous old sea clay soil, shallowly humose; parts of mapping units 62 and 63
Neth. '66	'Plas' earth soil (p.147 and 192); mapping unit Wo*l*

Location. Zuid-Holland Province, about 10 km north of Rotterdam (see Fig.1).
Parent material. Thin peaty lacustrine deposit overlying a thin remnant of the cut-over peat, overlying a fine-textured marine sediment (four to five thousand years old).
Topography and elevation. Level; on the floor of a former lake, reclaimed between 1836 and 1838; 5.3 m below sea level, about 3.5 m below the adjacent non-cut over peat land ('upland').
Drainage and ground water. No tile drains, open field drains spaced 10 m apart, ditches at intervals of 50-100 m. The water is pumped out of the polder into a so-called ring canal (the encircling canal which is roughly at the margin of the former lake), situated about 3.5 m higher, at the level of the upland. From there it flows into a minor lower branch of the Rhine, about 3 km to the southeast. The level of the ground water fluctuates between 30 and 70 cm.
Present land use. Grassland.
Range in land use. Practically everywhere used as grassland. When the peaty lacustrine material is thicker (then of course it is a Histosol), the soil is mostly used for horticulture: greenhouses with tomatoes and lettuce.

Discussion

Acid sulphate soils, originally called cat-clays by the Dutch farmers, were first studied in the Netherlands (Pons, in: Dost I, 1973, p.4-8). From the Symposium held on these soils in Wageningen in 1972, it is clear that they are widespread, especially on coastal lowlands in the tropics (Kawalec, in: Dost I, 1973, p.292).

The following properties are typical for the Dutch cat-clays: a peaty topsoil, a thin remnant of the cut-over peat (called peatlump by Edelman, 1950, p.164), a partly ripened cat-clay with yellow jarosite mottles and a 'bluish' non-ripened subsoil. In practice these soils are confined to the older marine sediments which outcrop on the floors of former lakes, where they have been revealed by peat cutting (Fig. 15, 16 and 17).

Classes defined intentionally for this kind of soil exist only in the recent systems of the FAO, England and Wales and the USA.

A low pH (less than 3.5) and the presence of the jarosite, are diagnostic of a sulfuric (sulphuric) horizon. In the British system such a horizon must have 'a pH of 3.5 or less in 0.01 M CaCl$_2$' (Avery, 1973), and in the FAO system as well in Soil Taxonomy a sulfuric horizon is characterized by a pH of less than 3.5 (H$_2$O, 1:1). The discussed soil has a pH of 3.4 (KCl) in its jarosite horizon; although this horizon is a classic cat-clay, it is on the boundary of the sulphuric horizon in these systems. Material that may change upon oxidation into a sulfuric horizon, i.e. potentially strongly acid material, is called sulfidic material in the FAO and USA systems: 'waterlogged mineral or organic soil materials that contain 0.75 percent or more sulfur (dry weight), mostly in the form of sulfides and that have less than three times as much carbonates (CaCO$_3$ equivalent) as sulfur' (SSS, 1975, p.63; FAO, 1974, p.31). In Soil Taxonomy these two diagnostic criteria give rise to several great groups and subgroups, but in the FAO system and the system of England and Wales there is only one class, Thionic Fluvisols and sulphuric humic alluvial gley soils respectively.

In Mückenhausen (1975, p.475) jarosite is mentioned in a *Moormarsch* (marine sediment overlying peat) and the local name in northern Germany is used: *Maibolt*. The presence of such material is indicated on the Soil Map of Niedersachsen, scale 1 : 25.000 (e.g. sheet 2310). *Pulvererde* (Benzler, in: Dost II, p.211-214) appears to contain more or less exactly the concept of sulfidic material.

This kind of soil is not mentioned in the first edition of the German system. The second edition (Mückenhausen et al., 1977, p.162 and 163) as well as Mückenhausen (1975, p.474) mentiones an analogous soil *Organomarsch*, however with a different genesis.

The material from the upper two horizons originally consisted of two components, an organic limnic sediment dating from the lake stage and clayey material originating from the reclamation (spoil from the newly dug ditches and the open field drains). This mixture has been homogenized into the A1 horizon, here divided into two subhorizons. The D layer is the undisturbed peat which first grew on this former tidal marsh, and was left after the peat cutting. The mineral subsoil is an old marine sediment. These lithological discontinuities could be indicated by the use of the Roman numeral prefixes II and III. Most probably the sulfuric horizon has to be coded B, so the horizon sequence according to the nomenclature of Soil Taxonomy might be: A11-A12-IIC-IIIBg-IIICg.

The G horizon is unripe, this is demonstrated in the laboratory data, by the high water content (127.2 g/100 g of dry soil) and in the field by the soft consistency (like soft soap) of the material which 'can be squeezed easily through the fingers' (De Bakker & Schelling, 1966, p.178). The C1g is a sulfuric horizon (low pH and jarosite mottles) and the G is sulfidic material in which sulphur is dominating over carbonate. Its pH is too low for reduced material, but this is an artifact introduced by routine treatment in the laboratory (airdried, ground etc.). Thus the pH measure is the oxidized pH value reflecting the change of the sulfidic material into material typical of a sulfuric horizon!

Profile description

A11	0-7 cm	Black (10YR2/1) peaty silty clay; moderate, fine subangular blocky structure; friable, slightly hard; abrupt, smooth boundary.
A12	7-23 cm	Black (10YR2/1) peaty silty clay; weak, compound coarse prisms, breaking to weak fine subangular blocky structure; friable, slightly hard; abrupt, smooth boundary.
D	23-30 cm	Very dark brown (7.5YR2/1.5) peat; oxidised and somewhat irreversibly dried, thus no botanical composition can be determined; the upper few cm consists of a greyish brown gyttja-layer; abrupt, smooth boundary.
C1g	30-50 cm	Grey (5Y4.5/1) silty clay; common to many, distinct medium pale yellow (5Y6.5/4) mottles (jarosite) and yellowish brown (10YR5/5) mottles, mainly alongside cracks and fossil reed-root channels; some cracks (one big one filled in with peat from the overlying horizon), but not yet cracked into big prisms, apedal; consistency terminology not applicable for this partly ripened somewhat slushy material.
CG	50-70 cm	Transitional horizon, both yellow and brown mottles decreasing with depth, and more slushy.
G	70-120 cm plus	Grey (10Y4.5/1) silty clay; no mottles; abundant remnants of reed; non-ripened, slushy material.

Analytical data

Horizon	Depth (cm)	Organic fraction				Particle-size distribution of mineral fraction (% of fine earth)				Field moisture (% of dry soil)
		O.m. (%)	C(%)	N(%)	C/N	< 2	2-16	16-50	> 50 μm	
A11	0-7	22.0	11.6	0.92	12.6	43	26	24	7	n.d.
A12	7-20	21.2	n.d.	n.d.	n.d.	47	25	23	5	n.d.
C1g	30-38	8.8	n.d.	n.d.	n.d.	56	27	16	1	n.d.
G	70-90	6.8	n.d.	n.d.	n.d.	57	28	15	0	127.2

Horizon	CaCO₃ (%)	pH-KCl	Extractable cations (% of sum)					meq./100 g			Ca/Mg	K-fix (%)
			Na	K	Mg	Ca	H	C.E.C.	CaO	SO₄		
A11	0.3	5.6	1.2	1.9	6.4	62.9	27.6	59.3	n.d.	n.d.	9.8	n.d.
A12	0.1	5.1	n.d.	n.d.	n.d.	n.d.	n.d.	n.d.	n.d.	n.d.	n.d.	n.d.
C1g	0.1	3.4	0.5	2.2	1.5	23.1	72.7	40.3	n.d.	n.d.	15.4	12
G	0.2	3.7	1.0	1.8	4.3	44.7	48.2	41.2	22	192	10.4	11

	0
A11	
	7
A12	
	23
D	
	30
C1g	
	50
CG	
	70
G	

Soils of the drained lakes and peat uplands district: LP3

General data

Classification

Eur.	Included in the Alluvial soils (p.4)
World	Mollic Gleysol (p.33), medium textured, level (p.5)
USA '38	Humic-Glei soil (p.119 in the 1949 modification)
USA '75	Fine-loamy, mixed (calcareous) mesic Fluvaquentic Haplaquoll (p.280)
Ger.	Not discussed, but analogous to the Tschernosemartiger Auenboden in the first edition of the German system (Mückenhausen et al., 1962, p.124 and 125) this soil could be classified as a Tschernosemartiger kalkhaltiger Seemarsch
Eng. & Wales	Calcareous humic-alluvial gley-soil
France	Sol hydromorphe peu humifère à gley profond (p.75)
Neth. '50	Calcareous old sea silt soil; mapping unit 1 (p.155-167)
Neth. '61	Calcareous old sea clay soil, shallowly humose, moderately sandy; mapping unit 55
Neth. '66	'Leek' earth soil (p.154, 193 and 194); mapping unit pMn55A

Location. Zuid-Holland Province, about 25 km southwest of Amsterdam (see Fig.1).

Parent material. Marine sediment, deposited four to five thousand years ago.

Topography and elevation. Level; on the floor of a former lake, reclaimed in 1768, about 5 m below sea level and about 3.5 m below the indisturbed peat land ('upland').

Drainage and ground water. The polder is drained by ditches spaced 125 m apart, and the field is tile drained, spaced 10-15 m apart. From the polder ditches the water is pumped in a ring canal (being roughly the boundary of the former lake) and from there into the main drainage system of western Holland. The ground-water table fluctuates between 35 cm and 1.10 m below the surface.

Present land use. Arable land, oat stubble.

Range in land use. Almost exclusively arable with main crops winter wheat, sugar beet and potatoes, with subsidiary colza, barley and peas. Because of the high organic matter content small grains and particularly flax are susceptible to lodging.

Discussion

Briefly this profile can be described as a non-saline, hydromorphic, calcareous soil developed from loamy, marine parent material, finely stratified at shallow depth with a dark topsoil rich in organic matter.

On the Soil Map of Europe it is clearly an Alluvial soil, but in the old USA system of 1938 there is no specific place for this soil. In the 1949 modification it fits more or less the definition of the Humic-Glei soil: 'An intrazonal group of poorly to very poorly drained hydromorphic soils with dark-coloured organi-mineral horizons of moderate thickness underlain by a mineral glei horizon' (Thorp & Smith, 1949, p.119).

Although there are slight differences between the definition of the mollic A horizon (FAO, 1974, p.24) and the mollic epipedon (SSS, 1975, p.14-16), the dark surface horizon of this profile meets the criteria for both. The soil has clear fluvic characteristics with no structural peds in the Cg horizon, which is finely stratified and has 1.1% organic matter.

In the FAO system there could be categories for a Mollic Fluvisol or a Gleyic Chernozem, but both are lacking. Fluvisols do not have a mollic A horizon and Mollic Gleysols precede Chernozems in the key and consequently Gleyic Chernozems cannot exist.

Soil Taxonomy takes into account both the hydromorphic and the fluvic characteristics, as well as the presence of a mollic epipedon in the name Fluvaquentic Haplaquoll.

The genesis of this Mollisol has nothing in common with the formation of the classic 'black earth' of the loess plains of central and eastern Europe. Those soils are developed under a steppe or a forest-steppe vegetation in a continental climate with a precipitation deficit, called by Němeček (1967, p.108) a 'non-leaching, periodically a leaching regime'. As a consequence, there are two periods every year with low biological activity, one in the cold winter and one in the dry summer. None of these conditions is present in the Netherlands. In fact, the dark topsoil results from a mixture of the organic sediment from the lake-stage (Cf. p.32) and the old marine material. The bulk of the organic matter is not derived directly from vegetation, but from this lacustrine mud.

There is little to discuss about the place of this soil in the systems of England & Wales and of France. In the former it is a calcareous humic-alluvial gley soil and in the latter the soil lies on the boundary between the *sols humiques à gley* and the *sols hydromorphes peu humifères*, 8% organic matter being the lower boundary of the first-mentioned *groupe* (CPCS, 1967, p.74).

If the above arguments are followed regarding the presence of two sediments, the use of Roman two for the lower two horizons (Ap1-Ap2-I/IIAC-IIC2g-IICG) could be advocated.

Disappearance of the stratification from the upper horizons of this soil (called homogenization in the Netherlands according to Hoeksema, 1953) is obviously an actual soil forming process. Based upon this evidence the C2g horizon might be called a CBg horizon according to the FAO legend (1974, p.22) and the CG horizon is a good example of a Gor horizon in the German system. The indication of such a transitional horizon between the Cg and the G horizons is not possible in most systems, and the modern USA system even amalgamated the old Cg and G (Cf. p.63 and 83).

In conclusion some remarks about the analytical data are necessary. The potassium fixation is low, as is usual in marine parent material (Cf. p.76 and 84). There might be a slight saline seepage or a remnant indication of the original situation as the percentage adsorbed magnesium in the subsoil is rather high (compare the recent marine soil M6, p.84).

Profile description

Ap1	0-20 cm	Black (10YR2.5/1) loam; structure typical for a plough layer in autumn: coarse clods resulting from the collapse of the seed bed, originally (after ploughing and harrowing) having a weak, very fine subangular blocky structure; friable; abrupt, smooth boundary.
Ap2	20-25 à 30 cm	Black (10YR2.5/1) loam, the lower boundary of this horizon indicates the depth to which it has been ploughed, in all cases this was not recently; moderate to strong, medium angular blocky structure; dense, few pores (plough-sole); firm to very firm; clear to gradual, smooth to irregular boundary.
ACg	25 à 30-50 cm	Transitional horizon with clear evidence of mixing A and C material by burrowing animals; few faint mottles; gradual, broken boundary.
C2g	50-110 cm	Grey (2.5Y5/1.5) very fine sandy loam (consisting of thin layers loam alternating with loamy very fine sand, only in the krotovinas these layers are mixed); common, distinct, medium mottles as brown (7.5YR5/4) streaks along former root channels; massive to single grain in the undisturbed part; weak very fine subangular blocky in the krotovinas; very friable; clear, smooth boundary.
C2G	110-120 cm plus	Dark grey (10Y4/1) loamy very fine sand, strongly stratified; common, distinct medium yellowish brown mottles (10YR5/5); transitional horizon to the G-horizon with neutral colours and no brown mottles.

Analytical data

Horizon	Depth (cm)	Organic fraction				Particle-size analysis of mineral fraction (% of fine earth)					
		O.m. (%)	C(%)	N(%)	C/N	< 2	2-16	16-50	50-105	105-150	> 150 μm
Ap1	0-20	7.5	3.6	0.34	10.6	23	12	24	20	11	11
Ap2	20-30	8.2	4.3	0.36	11.9	19	11	23	21	14	12
ACg	40-50	4.0	n.d.	n.d.	n.d.	23	15	26	21	10	4.5
C2g	55-70	1.1	n.d.	n.d.	n.d.	10	4.5	14	46	22	4.5

Horizon	CaCO₃ (%)	pH-KCl	Extractable cations (% of sum)					C.E.C. (meq. 100 g)	Ca/Mg	K-fix. (%)
			Na	K	Mg	Ca	H			
Ap1	1.8	7.0	1.4	1.4	2.2	91.4	3.6	36.2	41.5	n.d.
Ap2	2.3	6.9	1.6	0.5	3.1	90.6	4.2	38.1	29.2	n.d.
ACg	7.9	7.1	n.d.	n.d.	n.d.	n.d.	n.d.	n.d.	n.d.	n.d.
C2g	13.0	7.4	3.1	0.8	11.7	82.0	2.3	12.8	7.0	9

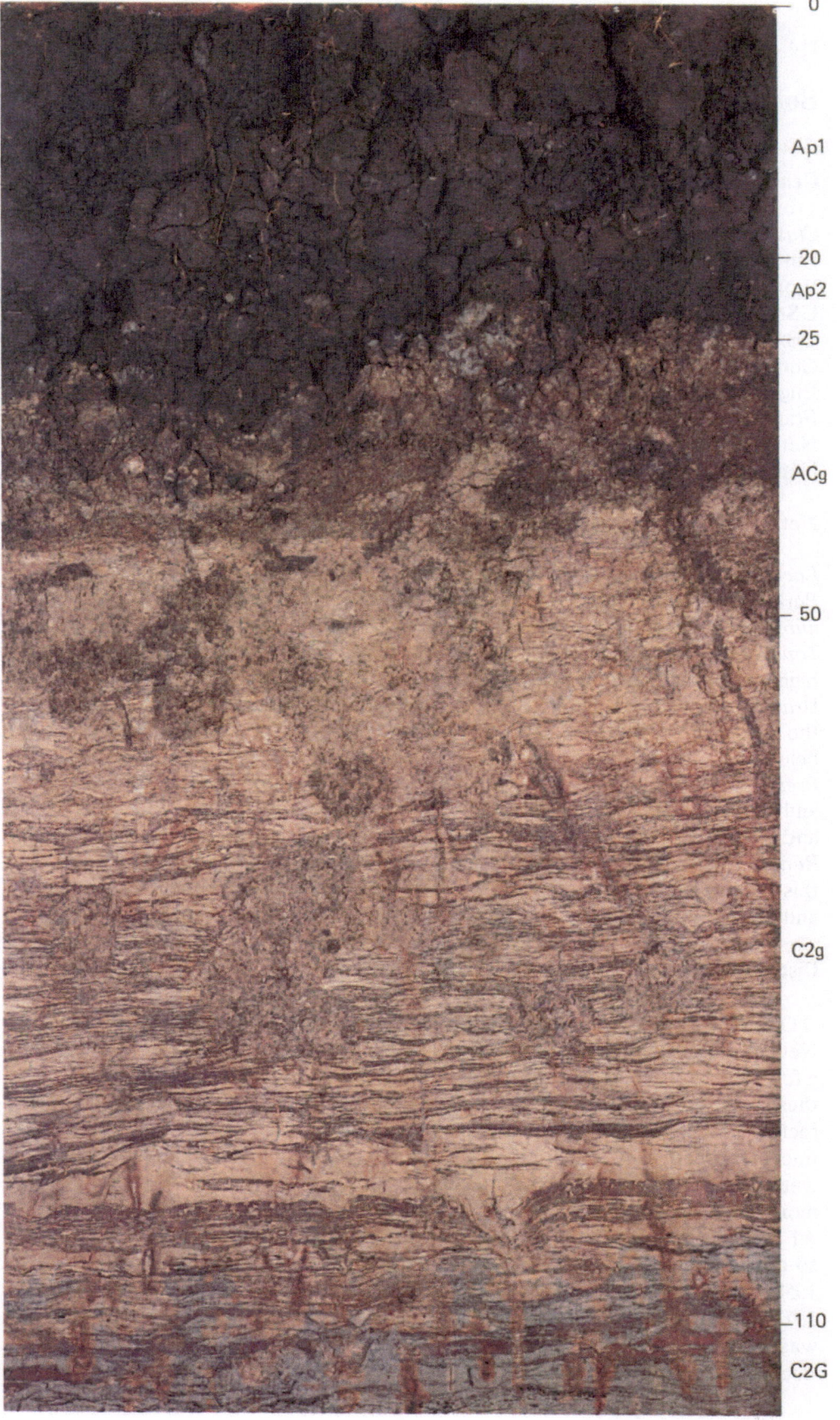

0

Ap1

20
Ap2

25

ACg

50

C2g

110

C2G

Soils of the cut-over raised bogs district: RB1

General data

Classification

Eur.	Organic soil (p.7), part of association O/P (p.78 and 79)
World	Dystric Histosol (p.41), level (p.5)
USA '38	Bog soil, Peat and Muck (p.1000 and 1128-1132)
USA '75	Dysic, mesic Typic Sphagnofibrist (p.216)
Ger.	Hochmoor aus Torfmoostorf (p.168 and 169)
Eng. & Wales	Raw oligo-fibrous peat soil
France	Sol à tourbe fibreuse, oligotrophe (p.74)
Neth. '50	Sphagnum peat soil; mapping unit 31 (p.78)
Neth. '61	High moor peat soil, not cut over, sphagnum moss peat, slightly or not oxidized; mapping unit 88
Neth. '66	'Vlier' peat soil (p.119 and 187); mapping unit Vsj

Location. In the southeast corner of Drente Province (see Fig.1).
Parent material. Oligotrophic peat (high moor peat), mainly derived from *Sphagnum* species.
Topography and elevation. Level; on islands of undisturbed peat, situated 2-3 m higher than the adjacent reclaimed soils; about 20 m above sea level.
Drainage and ground water. These remnants of a raised bog are drained because the lower lying arable land in the vicinity is also drained to a level of some metres below the surface of this soil.
Present land use. Waste land, dominant vegetation being ling (*Calluna vulgaris*) and cross-leaved heath (*Erica tetralix*), also purple moor-grass (*Molinia caerulea*) and some birches.
Range in land use. Today these peat remnants are waste, but up to the beginning of this century part of this area was cultivated after some drainage by shallow furrows and burning of the vegetation, buckwheat (*Fagopyrum esculentum*) was grown.

Discussion

Originally raised bogs overlying Pleistocene sands covered 150-200 000 ha of the Netherlands. The largest area was in the north of the country (Fig.21). Today only a few thousand hectares of raised bogs remain as waste. This does not imply that these areas are virgin; they are drained because of the drainage of the surrounding reclaimed cut-over areas. Actually growing untrafficable bogs are practically extinct in the Netherlands now (there are some small nature reserves, kept artificially wet in order to preserve a more or less natural site of living raised bogs). Furthermore there was a kind of slash-and-burn cultivation and perhaps the artefacts in the A1 horizon (see profile description) date from this time. In many places the so-called white peat or *bolster* (Dutch), *tourbe mousseuse* (Plaisance & Cailleux, 1958) or *tourbe blonde* (French), *Bleich-* or *Weissmoostorf* (Mückenhausen et al., 1977, p.168) partially has been removed for different purposes. It is known that it was used as bedding material in stables (Pape, 1970) and unlike plaggen manure, it decomposes completely, which may be the explanation for the lack of thick plaggen soils in the Pleistocene sandy district adjacent to the area of raised bogs. White peat was also (and is still) used when available for many horticultural purposes, such as for garden mulches. This material is a typical example of Fibric soil materials (SSS, 1975, p. 66); it is a young unweathered Sphagnum peat with clearly recognizable cell structure and because of this property it has a high and reversible water-holding capacity, seven or nine times its dry weight. Therefore is in excellently suited for horticultural purposes. Some *bolster* is still present between 10 and 48 cm depth, it might originally have been over one meter thick according to the local soil surveyor.

The lower horizons are called black peat (Casparie, 1972, p.144) by the peat-cutters (*zwartveen* -Dutch, *Schwarztorf* -German, *tourbe noire* -French). They consist of weathered peat which is much denser and has less favourable hydrological properties compared with the upper layers. It has a much lower and non-reversible water-holding capacity, which combined with its low ash content, means that this black peat is excellent fuel.

The classification of this soil yields no problems in the systems discussed; it is the *solum classicum* of a high moor developed from Sphagnum mosses.

All horizons have to be designated H in the FAO system; according to Soil Taxonomy (SSS, 1975, p.212) Oa-Oi-Oi-Oi and the lower horizons may be Oe; in the German system all horizons have to be Hh, the subscript h from *Hochmoor* = high moor; in the system of England and Wales: Oh-Of-Of-Of and lower down Om; in the French system the problem of horizon nomenclature in organic soils is not stipulated.

The Dutch system uses A1 for the earthified toplayer, which has been developed from the moss peat. The fresh non-weathered material is coded C2 and the lower peat layers which are slightly weathered and oxidized consequently could have been named C1, but it is preferred not to suggest genetic names for these layers, and they are merely numbered.

No laboratory data are given, except for the organic matter content of the moss peat which is always over 90-95%. The pH values are low (mostly below 3.5) and the C-N ratios are high (over 40). Instead, a brief botanical analysis reveals that the 'hairy' tussocks in the lowest horizon, visible on the plate, are remnants of cotton grass (*Eriophorum*), and the yellowish roots in the upper horizons are living roots of purple moor-grass (*Molinia caerulea*).

Profile description

A1   0-10 cm   Black (5YR2/1) mucky peat, strongly moulded material with few recogniza-
                        ble plant remains; some sand grains, fragments of glass and other resistant
                        material; abrupt, smooth boundary.
C21   10-28 cm   Reddish brown (5YR4/4) peat, non-moulded and non-weathered, coarse
                        peat; the so-called white peat or *bolster*; abrupt, smooth boundary.
C22   28-48 cm   Dark reddish brown (5YR3/3) peat, non-moulded and slightly weathered
                        les coarse peat; the so-called grey peat, i.e. *bolster* of a poorer quality than
                        required for horticultural purposes; clear, wavy boundary.
4, 5, 6
and 7   48-120 cm plus Dark reddish brown (5YR3/2), darker with depth, non-moulded and so-
                        mewhat weathered peat, the so-called black peat.

The bog floor is at 2 m depth, the sub-peat mineral soil below the peat is a hydromorphic podzol soil,
almost identical with soil P1, also developed in cover sand. All horizons and layers are non-calcareous.

Brief botanical data

A1   Seeds of *Chenopodiaceae*
C21   *Spagnum* spp., twigs of *Erica tetralix*
C22   *Sphagnum* spp., *Cyperaceae*
4   Many cyperaceous remains, e.g. *Carex* spp., many *Sphagnum* remains, e.g. *Sphagnum cuspi-
      datum* and *Hypnaceae,* seeds of *Menyanthes trifoliata*, leaves of *Andromeda polifolia* and some
      wing cases of beetles
5   Many cyperaceous and ericaceous remains, e.g. *Calluna vulgaris* and *Erica tetralix*, no *Sphag-
      num* remains
6   Many cyperaceous remains, many *Sphagnum* remains including some *Sphagnum cuspidatum*
      and remains of *Scheuchzeria palustris* and *Hypnaceae*
7   *Calluna vulgaris, Cyperaceae,* many *Eriophorum* fibres and remains of *Betula pubescens*
The layers 5 and 6 are called the hummock-and hollow-system by Casparie (1972, p. 190), a hummock is
observable on the plate (left, middle part).

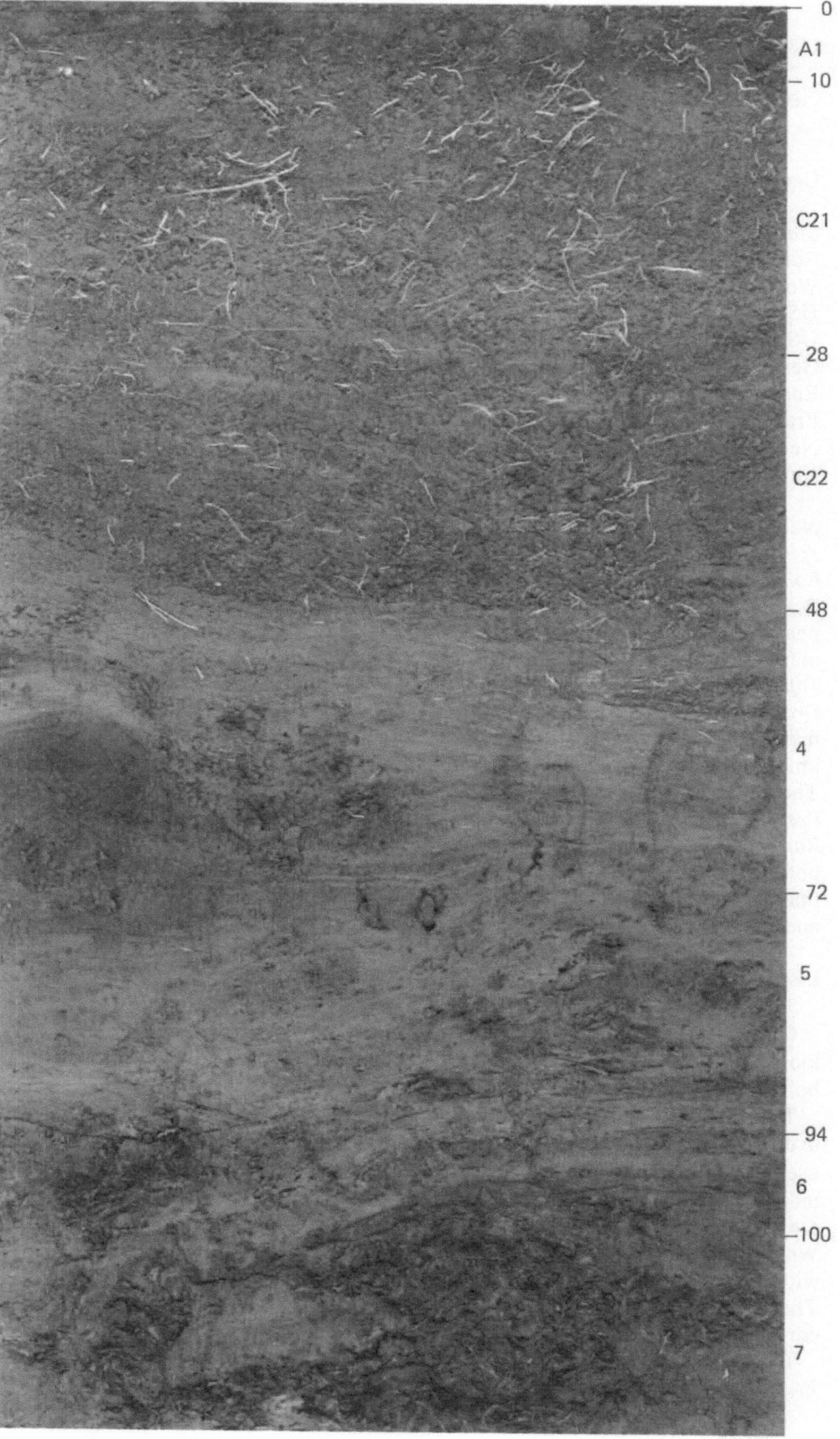

	0
A1	10
C21	
	28
C22	
	48
4	
	72
5	
	94
6	100
7	

Soils of the cut-over raised bogs district: RB2

General data

Classification

Eur.	Organic soil (p. 7), part of association O/P (p. 78 and 79)
World	Dystric Histosol (p. 41), level (p. 5)
USA '38	Bog soil, Peat and Muck (p. 1000 and 1128-1132)
USA '75	Dysic, mesic Plaggeptic (?) Sphagnofibrist (p. 216)
Ger.	Sanddeckkultur (p. 169)
Eng. & Wales	Earthy oligo-fibrous peat soil
France	Sol à tourbe fibreuse, oligotrophe (p. 74)
Neth. '50	Reclaimed peat sub-soil; mapping unit 37 (p. 79-83)
Neth. '61	High moor peat reclamation soil, younger soil reclaimed from cut-over peat; mapping unit 96
Neth. '66	'Meer' peat soil (p. 118 and 187); mapping unit iVz

Location. In the southeast corner of Drente Province (see Fig. 1).

Parent material. Oligotrophic peat, mainly *Sphagnum,* overlying mesotrophic fen peat, Pleistocene sandy subsoil at a depth of 115 cm. The sandy topsoil is added by man.

Topography and elevation. Level; about 14 m above sea level.

Drainage and ground water. The site is drained by ditches spaced 100 m apart into main canals, which discharge into the sea by means of a pumping station about 100 km to the north. Between the sampling site and the sea are many locks in the canals. The level of the ground water fluctuates between 30 and 60 cm below the surface.

Present land use. Arable land, potatoes.

Range in land use. Almost exclusively arable, the most important crop is starch potatoes, followed by small spring grains (wheat, oats and barley), some sugar beet (increasing in the last years), few winter crops like winter wheat, barley and rye, and some miscellaneous crops like beans and peas for canneries.

Discussion

On this site peat cutting has removed some metres of peat, and part of the upper, loose, spongy young Sphagnum peat (cf. p. 115) has been put back (the Dp horizon); the layers below 25 cm depth are undisturbed.

The spoil derived from digging the shipping canals and the ditches has been used to dress the cut-over land with 10-20 cm of sand (Aanp.) Non-sanded peat soils when used as arable land hold some disadvantages over their sanded counterparts: the more organic the topsoil the less is its bearing capacity, it is more prone to wind erosion and the effect of the late spring frosts on potatoes are more devastating. When walking on the humose sandy topsoil of arable fields, a surveyor unfamiliair with the history of this peat reclamation area would expect to find mineral soils. The result of a boring or the presence of lumps in peat in freshly ploughed land corrects this first false impression.

The system of England and Wales refers to peat soils with a mineral surface layer, but, unlike the Dutch system, peat soils having an earthy topsoil are combined with those having a mineral surface layer into one group, the earthy peat soils. At subgroup level this class is only subdivided according to different kinds of peat.

In the first edition of the German system there was no specific class for this soil. The recent book by Mückenhausen (1975, p. 479) contains a description of similar soils within the *Anthropogene Moore* as *Sanddeckkultur,* the same name is used in the Arbeitsgemeinschaft Bodenkunde (1971, p. 130) and in the second edition of the German system (Mückenhausen et al., 1977, p. 169). In the other systems discussed sanded peat soils are not mentioned. In Soil Taxonomy there is also no special subgroup for this soil, it could be named Plaggeptic Sphagnofibrist.

The mineral part of the Aanp is derived from the sub-peat mineral soil profile, not from the same site but from the spoil of the ditches and canals surrounding the fields (see above and p. 39). Its organic content is derived partly from decomposed peat mixed in by cultivation, and partly from the more normal sources of organic matter in a plough layer, decaying remnants of the crops and organic manure.

The academic question arises (for an analogous discussion see p. 107), about how best to use figures as prefixes to mark lithological discontinuities in this profile description. There are three strata: the man-made plough layer, the peat and the Pleistocene sand, making it a I-II-III-soil. However, to stress the fact that the surface sand (Aanp) is derived from the substratum, there could be two strata, designated I-II-I (or even I/II-II-I, to indicate that the plough layer is a mixture of sand and decomposed peat). In the Dutch edition the old-fashioned concept of D is used, and the buried soil labelled with the suffix b. The reworked upper peat layer has the suffix p and the two lowest reduced peat layers the suffix G.

No analytical data are given for this soil, but the botanical data are interwoven in the profile description.

The evaluation of the organic matter content of these soils (with reference to liming and fertilizing requirements) is difficult, because part of the organic matter is humus and part is inert as non- or partly decomposed peat.

The organic matter content of the Aanp is seldom below 8% or over 25%, usually it is between 12 and 18%. This depends on the thickness of the Aanp which varies with the distance from the source of the spoil used to dress the peat. Thinner Aanp horizons have a higher organic matter content.

Profile description

Aanp	0-12 cm	Dark grey (10YR4/1) fine sand; single grain; loose; abrupt, smooth boundary.
Dp	12-25 cm	Dark reddish brown (5YR3/3) peat; reworked young *Sphagnum* peat; loose spongy material; abrupt, smooth boundary.
D2	25-45 cm	Dark reddish brown (5YR2/2) peat; undisturbed old *Sphagnum* peat of a more close-packed structure and a denser consistency than the overlying young *Sphagnum* peat; remnants of ling (*Calluna vulgaris*) and cottongrass (*Eriophorum*), at the boundary with the underlying horizon, which is abrupt and smooth, remnants of Scots pines and birches.
DG1	45-66 cm	Dark reddish brown (5YR3/2) peat, which changes upon oxidation to a darker shade; dense peat with remnants of *Scheuchzeria;* abrupt, smooth boundary.
DG2	66-105 cm	Dark reddish brown (5YR3/4) peat, colour change upon oxidation as above; dense peat with remnants of birch and oak, sedges, some reed (?) (*Phragmites communis*) and reedmace (*Typha*); abrupt smooth boundary.
A1b	105-115 cm	Black (5YR2/1), mucky fine sand; slightly sticky, slightly plastic; some roots of reed, partly filled in with amorphous humus: dopplerite; abrupt, smooth boundary.
Cb	115-120 cm plus	Pale brown (10YR6/3) fine sand; single grain, uncoated sand. All horizons and layers are non-calcareous.

	0
Aanp	
	12
Dp	
	25
D2	
	45
DG1	
	66
DG2	
	105
A1b	
	115
Cb	

Soils of the cut-over raised bogs district: RB3

General data

Classification

Eur.   No suitable category in the list of soil units; on the map included in the association of organic soils and podzolized soils, symbol O/P (p. 78 and 79)

World   No suitable category

USA '38   No suitable category

USA '75   Histic over sandy, mesic Histic (?) Haplaquod (p. 336 and 337)

Ger.   No suitable category

Eng. & Wales No suitable category

France   No suitable category

Neth. '50   Included in the reclaimed peat sub-soils; mapping unit 37 (p. 79-83)

Neth. '61   High moor peat reclamation soil, medium high, older soil from cut-over peat, sand and peat; mapping unit 98

Neth. '66   'Dam' podzol soil (p. 125 and 188); mapping unit iWp

Location. Groningen Province (see Fig. 1). Together with soil RB4 (p. 126) sampled on an experimental farm, where different methods of subsoiling are tested in practice.

Parent material. Humose sand overlying a thin remnant of cut-over peat, overlying a buried podzol in Pleistocene sand, viz. cover sand.

Topography and elevation. Level; about 2 m above sea level.

Drainage and ground water. Drained by ditches spaced 75 m apart, not tile drained; ground-water level between 50 cm and 1.20 m.

Present land use. Leys.

Range in land use. This soil is surrounded by soils like RB2 and its crop rotation therefore does not differ very much from soil RB2 (p. 118).

Discussion

The initial history of each of these cut-over raised bog soils is similar: some metres of peat have been removed by cutting for fuel and the remnants of peat have been sanded with spoil dug from the ditches (see also p. 118); these created 'soils' are used predominantly as arable land.

Differences in thickness of the left-over peat (compare this soil with the previous one) are due to the slightly undulating bog floor (Fig. 20). Before being blanketed by the bog, podzols developed in the sandy substratum above the watertable, whereas lower places showed little soil formation. This soil was sampled at a site with a thin remnant of peat and a buried hydromorphic podzol at shallow depth.

Except in the old USA system, all systems discussed have defined boundaries between organic and mineral soils. These definitions are based mainly on thickness and organic matter content of the organic layer.

The data in Table 6 are somewhat generalized (e.g. Avery uses the boundary of more than 30% o.m. if the 'inorganic fraction is 50 per cent of more clay', instead of 60% clay in the FAO system and Soil Taxonomy). The German data are partly from

Table 6. Main items of the definitions of organic soils in the systems discussed

System	Thickness	O.m. content
Eur.	> 30 cm	> 30%
World	> 40 cm	> 20% if no clay
	(> 60 cm if Sphagnum)	> 30% if > 60% clay
USA '75	like World	like World
Ger.	> 30 cm	> 30%
	> 20 cm	
Eng. & Wales	> 40 cm	like World

Mückenhausen et al. (1977, p. 167: 20 cm) who indicated no limitation for organic matter content, and from Arbeitsgemeinschaft Bodenkunde (1971, p. 42). As appears from Table 6 there are slight differences between the systems discussed, but in each system this is classified as a mineral soil.

The thin peat layer satisfies the definition of the histic epipedon (SSS, 1975, p. 17, item 3) in that 'A layer of organic material that. . . lies beneath a surface layer of mineral materials that is < 40 cm (16 in.) thick'. In the definition of the histic H horizon (FAO, 1974, p. 24) no reference is made to shallow buried, thin organic layers. As a consequence, the adjective Histic only can be used in classification according to Soil Taxonomy.

In the Federal Republic of Germany (Arbeitsgemeinschaft Bodenkunde, 1971, p. 116) such soils could be called *Hochmoor-Podsole,* or an *Anmoorgley-Podsol* (Mückenhausen et al., 1977, p. 112; Mückenhausen, 1975, p. 448), but no reference is made to a mineral topsoil.

The plough layer of this soil is coded Aanp; the thin layer of peat D and the horizons of the buried podzol have the suffix b. Arguments about the use of these codes can be found in the analogous discussion about soil RB2 (p. 119).

The C-N ratio of the plough layer is rather high, most probably this is caused by the mixture of normal organic matter and partly decomposed peat in the Aanp which has a C-N ratio about 50.

As can be expected from its origin, the particle-size distribution of the topsoil resembles the texture of the substratum.

Profile description

Aanp	0-13 cm	Black (7.5YR2/1) loamy fine sand; single grain; loose to very friable; abrupt, smooth boundary.
D	13-26 cm	Black (5YR2/1) strongly and very finely laminated oligotrophic peat, few recognisable remains like *Calluna* and *Scheuchzeria;* abrupt, smooth boundary.
A0b	26-30 cm	Black (5R1.5/1) sandy muck; the original sod of the overgrown (buried) podzol; greasy material; abrupt, smooth boundary.
B2b	30-60 cm	Dark brown (7.5YR3/4) fine sand; stained with coatings of amorphous organic matter, no iron coatings; common, prominent, fine mottles being vertical streaks filled in with organic matter, obviously a fossil root system; at 50 cm depth a very thin lamella with much amorphous organic matter; massive to single grain; very friable; diffuse, smooth boundary.
B3b	60-120 cm plus	Dark yellowish brown (10YR4/4) fine sand, characterized by a decrease in number of the organic mottles, and thinner coatings of organic matter on the sandgrains, but like soil P 1 the sandgrains are stained with brownish organic matter to a considerable depth.

All horizons and layers are non-calcareous

Analytical data

Hori-zon	Depth (cm)	pH-KCl	Organic fraction				Particle-size distribution of mineral fraction (% of fine earth)					
			O.m. (%)	C(%)	N(%)	C/N	< 2	2-16	16-50	50-105	105-150	> 150 μm
Aanp	0-13	4.7	10.5	6.1	0.30	20.3	6	2.5	9	21	32	30
D	13-26	3.6	83.4	n.d.	n.d.	n.d.	n.d.	n.d.	n.d.	n.d.	n.d.	n.d.
B2b	35-50	3.9	2.6	1.5	0.07	21.4	3.5	0.5	6	29	39	23
B3b	100-110	4.5	0.4	n.d.	n.d.	n.d.	4	0	7.5	27	34	27.5

Detailed particle-size distribution of the sand fraction of the B3 sample:

50-75	75-105	105-150	150-210	210-300	> 300 μm
10	17	34	18	7	2.5

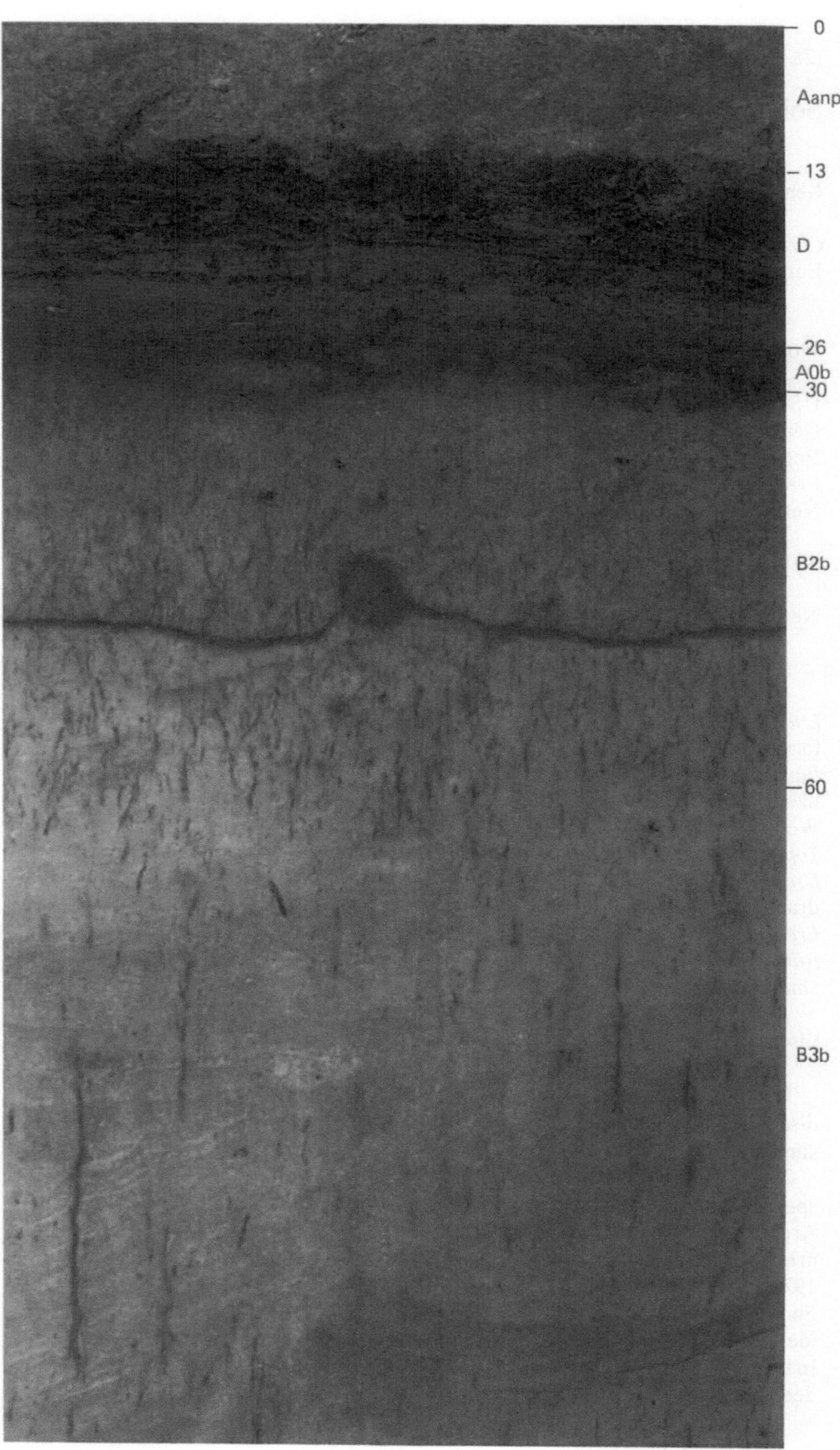

0

Aanp

— 13

D

— 26
A0b
— 30

B2b

— 60

B3b

Soils of the cut-over raised bogs district: RB4

General data

Classification

Eur.	No suitable category in the list of soil units, on the map included in the association of organic soils and podzolized soils, symbol O/P (p. 78 and 79)
World	No suitable category
USA '38	No suitable category
USA '75	Sandy, mesic Haplaquodic (?) Arent (p. 187)
Ger.	Sandmischkultur (p. 170)
Eng. & Wales	Disturbed soil
France	No suitable category
Neth. '50	Included in the reclaimed peat sub-soils; mapping unit 37, (p. 79-83)
Neth. '61	Not mentioned as such, either on the map or in the legend; part of the same unit as soil RB3 (p. 122)
Neth. '66	Like soil RB3 (p. 122), but this deeply ploughed soil gets an additional symbol behind the code of the mapping unit to indicate this deep disturbance

Location. Groningen Province (see Fig. 1). Sampled on the same experimental farm as soil RB3 (p. 126), but on an adjoining field.
Parent material. Material mixed by deep ploughing (humose sand, peat and podzolized sand, see soil RB3, p. 122) overlying undisturbed sand (aeolian sand from the Weichsel Age).
Topography and elevation. Level; about 2 m above sea level.
Drainage and ground water. Drained by ditches spaced 75 m apart, not tile drained; ground-water level between 50 cm and 1.20 m.
Present land use. Fallow before sowing sugar beet.
Range in land use. Surrounded by soils like soil RB2 (p. 118) its crop rotation is similar.

Discussion

This discussion, and also the discussion about soil RB5, p. 131, parallels the discussion about soil RB3, p. 123, because before deep ploughing these soils were similar.

Only according to Soil Taxonomy, the German and the British systems are there special classes for such reworked soils. In the new USA system they are called Arents, 'Entisols that have fragments of diagnostic horizons between 25 cm and 1 m below the surface, but the fragments are not arranged in discernible order' (SSS, 1975, p. 180). 'Subgroups of Arents are intergrades to suborders or great groups of Spodosols, . . . or other orders, according to the nature of the fragments that can be identified. . .' (SSS, 1975, p. 187), therefore they are named Haplaquodic Arents. In the German system they could be Rigosols: 'But we want to place soils turned by deep ploughing also in the type of Rigosols' (Mückenhausen et al., 1977, p. 140), but this is clearly not the intention, they must be called *Sandmischkultur* (Hugen-roth, 1971, p. 132, 133 and 148; Mückenhausen, 1975, p. 479; Mückenhausen et al., 1977, p. 170). The system of England & Wales has provided a special group for disturbed soils in which 'deeply worked ground' is included (Avery, 1973).

Upon drainage, organic soils contract because of water loss and compaction and partly because of wastage that is accelerated when the soils are used as arable land. Seale (1975, p. 77, 78 and 172) mentions an average wastage of 1.8 cm a year in the English Fenlands north of Cambridge. Mückenhausen (1975, p. 479) mentions 1 cm a year, both authors refer to peat soils with an organic plough layer. For the Dutch soils of the cut-over raised bogs with a sandy topsoil, Booij (1959) also estimates a loss of 1 cm a year, which he ascribes to the oxidation of the peat ploughed into the sandy topsoil (for further discussion about the reworked soils see p. 130 and 131).

The plate shows a section through a furrow, i.e. perpendicular to the direction of ploughing. No satisfactory method has been developed yet to describe such mixed soils, so instead of a normal profile description, a sketch-section is given opposite the plate using the same horizon nomenclature for the fragments as was used for the horizons of soil RB3 (p. 125).

Fig. 36. The plough used to plough soil RB4 to a depth of 1.10 m.

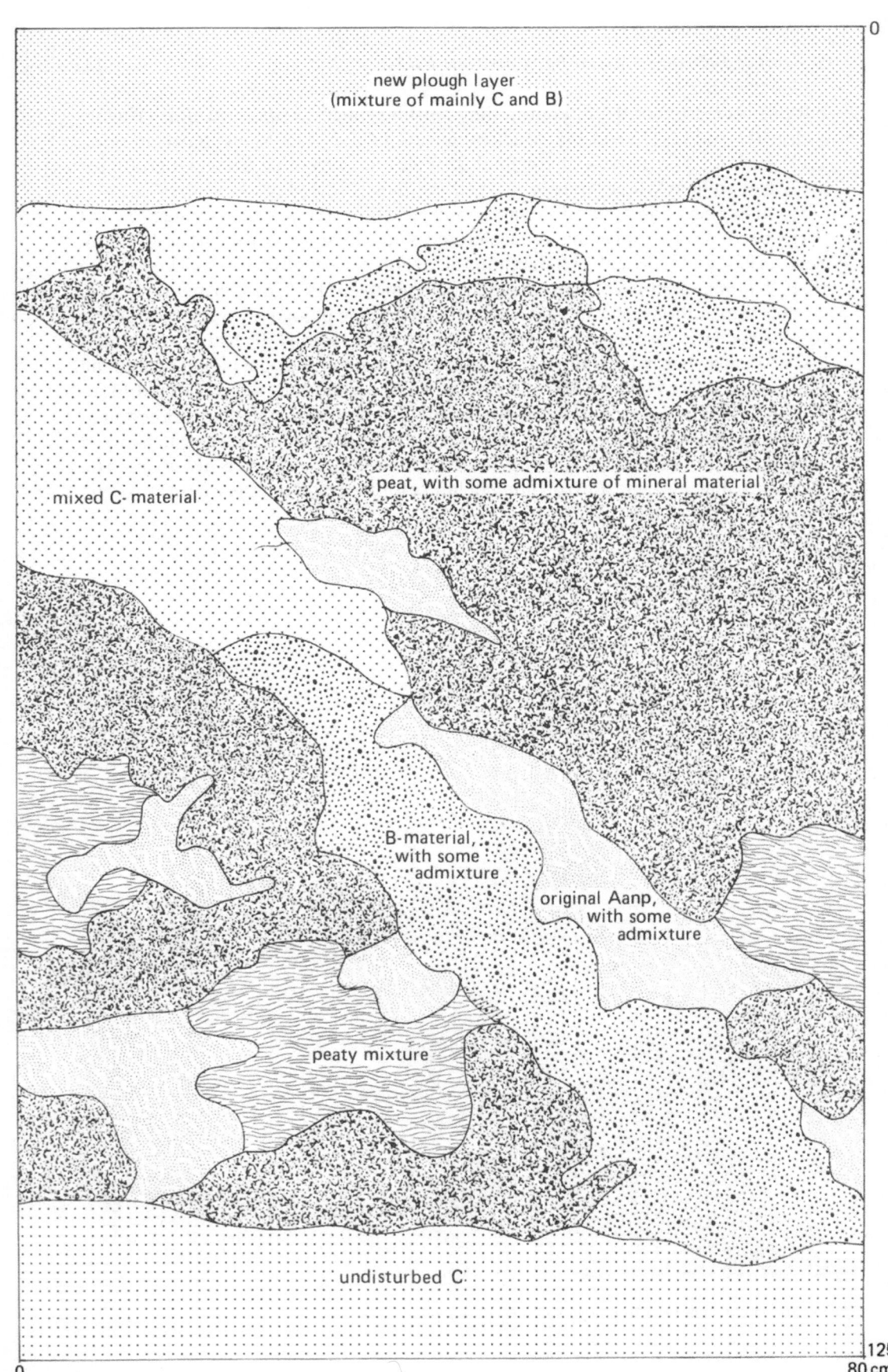

Soils of the cut-over raised bogs district: RB5

General data

Classification

Eur.   No suitable category in the list of soil units, on the map included in
               the association of organic soils and podzolized soils, symbol O/P
               (p. 78 and 79)
World   No suitable category
USA '38   No suitable category
USA '75   Sandy, mesic Haplaquodic (?) Arent (p. 187)
Ger.   Sandmischkultur (p. 170)
Eng. & Wales   Disturbed soil
France   No suitable category
Neth. '50   Included in the reclaimed peat sub-soils; mapping unit 37 (p. 79-83)
Neth. '61   Not mentioned as such, either on the map or in the legend; part of
               the same mapping unit as soil RB3 (p. 122)
Neth. '66   Like soil RB3 (p. 122). This soil has an additional symbol behind
               the code of the mapping unit to indicate deep disturbance by
               subsoiling

Location. In the north of Overyssel Province (see Fig. 1).
Parent material. Material mixed by subsoiling (humose sand, peat and podzolized
sand, see soil RB3, p. 122) overlying undisturbed sand (aeolian sand of Weichsel
Age).
Topography and elevation. Level; about 10 m above sea level.
Drainage and ground water. Drained by ditches spaced 80 m apart, not tile
drained; ground-water level between 0.7 and 1.5 m.
Present land use. Arable, freshly sown barley.
Range in land use. This soil is surrounded by soils like soil RB2 (p. 118) and its
crop rotation is similar.

Discussion

There are several arguments in favour of mixing soils with a thin peat layer
overlying a podzol, either by deep ploughing or by subsoiling. Firstly, the buried B
horizon which is often rather firm or even indurated, especially when lying immediately above ground-water level. Secondly, the remnant of black peat (see p. 115)
often has a low permeability and a high acidity. Both features can obstruct root
development and the first-mentioned property causes temporary waterlogging.
Disrupting and mixing these layers will diminish the effect of these unfavourable
properties. Thirdly, as a result of the slightly undulating bog floor (Fig. 20, p. 36)
the ground level after the original reclamation becomes uneven as a result of
differential shrinkage, thus causing wet spots locally. Soil amelioration includes
subsoiling and levelling the field, thus improving both faults at the same time.
Fourthly, the plough layer is often too humose; crops on such fields suffer more
from late frosts, and weed control is difficult.
On a pilot farm of the local Farmers Association different methods of loosening

the subsurface layers have been tried. Soils have been deeply ploughed with and without (soil RB4) preservation of the original topsoil. Other fields have been subsoiled with different kinds of implements; one field had an additional topdressing of 8 cm sand and another was kept in its original state (soil RB3). This soil is not from the pilot farm, but from a field of another farm where a new kind of subsoiler has been tried. The soil with the new topsoil (soil RB4) needs more fertilizers, especially nitrogen but all mixed soils give higher yields than the original soils.

This soil has been described with help of a sketch-section also; unlike all the other plates, the depth of this soil is shown perpendicular to the spine of the book.

Fig. 37. The mixture of the subsoil of soil RB5 was obtained by using this subsoiler.

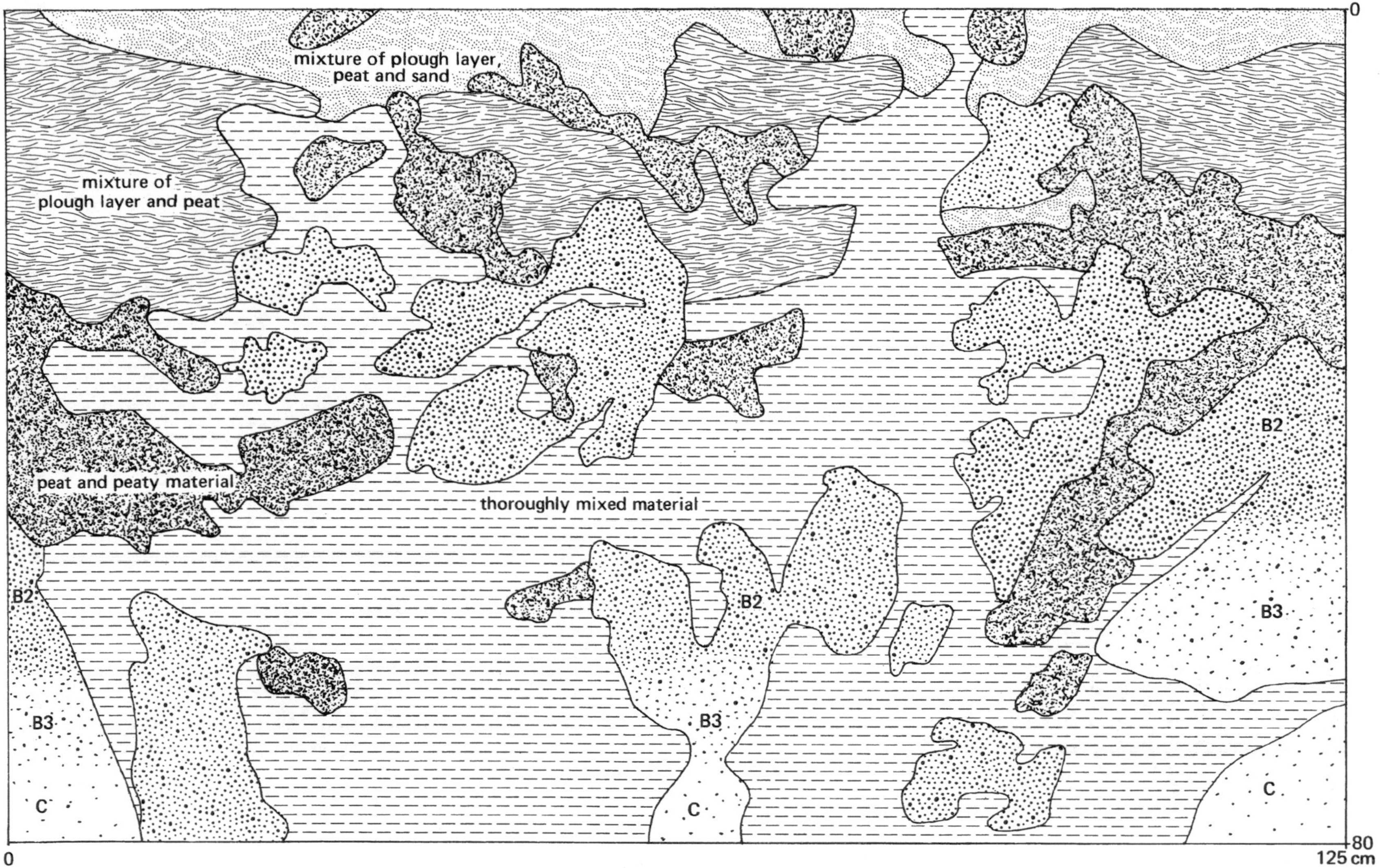

mixture of plough layer, peat and sand

mixture of plough layer and peat

peat and peaty material

thoroughly mixed material

B2

B3

C

0

80

125 cm

0

Anthropogenic soils: A1

General data

Classification

Eur.	No suitable category
World	No suitable category
USA '38	No suitable category, either in the original list of classification units, or in the 1949 modification
USA '75	Sandy, siliceous, mesic Plaggept (p. 257)
Ger.	Brauner Podsol-Plaggenesch (p. 138)
Eng. & Wales	Sandy man-made humus soil
France	Sol d'apport anthropique (p. 25)
Neth. '50	Loamy sand soils, old arable land; mapping unit 43 (p. 23 and 24)
Neth. '61	Non-calcareous sand soil, high old arable land, very loamy fine sand, with brown topsoil; mapping unit 115k (in fact sampled within soil association 131)
Neth. '66	Brown 'enk' earth soil (p. 144 and 191); mapping unit bEZ23

Location. Gelderland Province, about 50 km northeast of Arnhem (see Fig. 1).
Parent material. Cover sand, an aeolian deposit of Weichsel age.
Topography and elevation. Situated on a ridge, a few kilometres long and 50-100 m wide, about 3-4 m above the neighbouring brook valley and about 18 m above sea level.
Drainage and ground water. No external drainage, no runoff due to the rapid permeability; the ground-water level is always deeper than 3 m.
Present land use. Arable land, rye stubble.
Range in land use. Until recently it was almost exclusively arable. Its genesis resulted from an age-long use of earth-containing manure on arable land, hence the name 'old arable land' on earlier Dutch soil maps. Dominant crops were rye, oats, potatoes and mangolds, after rye Swedish turnips were often grown as a second crop. As in other regions of the Pleistocene sandy district, land use has shifted from exclusively arable to predominantly grass. Of the remaining arable crops silage-maize has become very important; there is also some wheat and sugar beet, and in the southern Netherlands some orchards and nurseries (roses and shrubberies).

Discussion

The brown and black plaggen soils (soils A1 and A2) have much in common, so they will be discussed together, beginning on this page and continuing on p. 138.

The upper two horizons of these soils consist of allochthonous material, gradually added by man in the practice of using earth-containing manure. In the Pleistocene sandy district turves (= *plaggen,* both in Dutch and in German) were used as bedding materials in the stables. The mineral content of the manure is derived from these turves. The farmers were probably unaware of the fact that they were fertilizing their soils and also were very gradually raising them. Dr. W. C. H. Staring (1808-1877) a geologist by profession but also one of the first Dutch soil scientists, wrote in 1856 in the first *De Bodem van Nederland* (Soils of the

Netherlands): '. . . assuming that one hectare of land is manured every three years with 80 cart-loads of plaggen manure, leaving 40 cubic metres of dark soil after decay, the soil would be raised 4 millimetres, and the raising of one metre would require 750 years'.

The plaggen soils are discussed in a review by Conry (1974), but this discussion includes soils other than classic plaggen soils of the Pleistocene sandy district of northwestern Europe, as indicated by the subtitle of this article: 'A review of man-made raised soils'. It includes the Irish sanded soils which have been raised by carting calcareous beach sand to acid autochthonous soils, but it also includes profiles such as soil A4 in this book (p. 146-149).

The plaggen soils sensu stricto occur on the Pleistocene sandy part of the mainland of northwestern Europe (Fig. 25) where they may be found in individual areas too small to be delineated on small-scale soil maps. They do not appear on the Soil Map of Europe, scale 1:2 500 000, but are mentioned in the Explanatory Text as inclusions in the association of podzolized soils; 'These soils, called *Plaggengrond* in Belgium and the Netherlands or *Plaggenböden* in Germany, are generally extensive around villages or old homesteads' (Dudal et al., 1966, p. 46 and 47). They are not actually delineated on the Soil Map of the Federal Republic of Germany, scale 1:1 000 000 (Hollstein, 1963), but the eastern and southern limit of their occurence is indicated, a line marked on Figure 25). Near the western boundary of the German Democratic Republic there are some small areas of *Plaggenböden* (personal communication Prof. Dr G. Haase).

In the Legend to the Soil Map of the World there is no suitable category for plaggen soils, but they are mentioned in the text: '. . . the scale of the map did not allow for separating soils which are characterized by a plaggen epipedon' (FAO, 1974, p. 23).

In the German system these soils are discussed at length (Mückenhausen et al., 1977, p. 137). There is a special *Klasse* for *Terrestrische anthropogene Böden,* subdivided into three *Typen,* one of them being the *Plaggenesch.* Both A1 + A2 soils fit into this class; at the level of the *Subtyp* soil A1 is a *brauner Podsol-Plaggenesch* and soil A2 a *grauer Podsol-Plaggenesch.* Two differentiating kinds of criteria are used at this level; the kind of buried soil (in both cases a *Podsol,* but there exists a *grauer Gley-Plaggenesch*) and the colour of the Plaggen epipedon. In the German system there is a difference between brown and grey plaggen soils, the latter being labelled black in the Dutch system (De Bakker & Schelling, 1966. p. 174, 175 and 191).

Avery (1973) defines a major group of man-made soils for England and Wales; subdivided into two groups (man-made humus soils and disturbed soils). The first group must have 'a thick man-made A horizon', according to the definition, with a dark horizon more than 40 cm thick, resulting from addition of earth-containing manure or otherwise attributable to human activity. This group has two subgroups, one sandy and one earthy (earthified peat in the Dutch system); both soils A1 + A2 clearly fit into the sandy subgroup. Avery provides a subgroup for a kind of soil only recently described in Cornwall on Sheet SW53 (according to the Report for 1974 of Rothamsted Experimental Station).

Plaggen soils are not known to occur in France, but there exists a *groupe des sols*

(To be continued on p. 138)

Profile description

Aanp	0-23 cm	Very dark grey brown (10YR3/2) loamy fine sand; single grain to very weak fine subangular blocky structure; very friable, soft; abrupt, smooth boundary.
Aan2	23-90 cm	Dark brown (7.5YR3/3) fine sandy loam; very weak, fine subangular blocky structure; very friable to friable, soft; few to common, medium clods of grey, mottled sandy loam (not recognisable on the plate), wormholes partly filled in with Aanp-material; small fragments of bricks, charcoal, pottery and the like; abrupt, smooth boundary.
Apb	90-115 cm	Dark grey (10YR4/1) fine sand; single grain; loose; obviously a buried plough layer; abrupt and somewhat irregular boundary.
Bb	115-120 cm	Yellowish brown (10YR5/5) fine sand; single grain; loose; the yellowish colour is due to thin iron coatings on the sand grains.

The soil is non-calcareous throughout.

Analytical data

Hori- zon	Depth (cm)	pH- KC1	Organic fraction				Particle-size distribution of mineral fraction (% of fine earth)					
			O.m. (%)	C(%)	N(%)	C/N	< 2	2-16	16-50	50-105	105-150	> 150 μm
Aanp	0-20	4.5	4.3	1.7	0.19	8.9	7	2.5	11	19	22	38
Aan2	45-60	3.8	3.7	1.5	0.14	10.7	10	4	12	17	19	38
Cb	120-125	4.7	0.8	n.d.	n.d.	n.d.	1	0.2	2.5	12	20	64.9

Detailed particle-size distribution of the > 150 μm fraction of the Cb sample:

150-210	210-300	300-420	420-600	600-850	850-1200 μ m
29	24	10	1.5	0.3	0.1

0

Aanp

— 23

Aan2

— 90

A1pb

—115

Bb

Anthropogenic soils: A2

General data

Classification

Eur.	No suitable category
World	No suitable category
USA '38	No suitable category, either in the original list of classification units (1938), or in the 1949 modification
USA '75	Sandy, siliceous, mesic Plaggept (p. 257)
Ger.	Grauer Podsol-Plaggenesch (p. 138)
Eng. & Wales	Sandy man-made humus soil
France	Sol d'apport anthropique (p. 25)
Neth. '50	Sandy soil, old arable land; mapping unit 45 (p. 23 and 24)
Neth. '61	Non-calcareous sand soil, high old arable land, slightly loamy fine sand; mapping unit 114
Neth. '66	Black 'enk' earth soil (p. 144, 145 and 191); mapping unit zEZ21

Location. Gelderland Province, about 50 km northeast of Arnhem (see Fig. 1).
Parent material. Cover sand, i.e. an aeolian deposit of Weichsel age.
Topography and elevation. Situated on a small ridge, a few hundreds of metres long and about 50 m wide, elevated about 2 m above the surrounding fields; about 23 m above sea level.
Drainage and ground water. No external drainage, no runoff due to the rapid permeability; ground-water level always deeper than 2 m.
Present land use. Arable land, potatoes.
Range in land use. Until recently it was almost exclusively arable. The crop rotation was narrower than of the preceding soil A1, because less moisture can be provided due to the lower clay content of this soil. There was no wheat, no sugar beet and no horticulture on this black plaggen soil. In the last decade the area of grassland, especially long-term leys, has increased considerably, with silage-maize replacing mangolds and Swedish turnips as a fodder crop.

(Continuation from p. 135)

anthropiques in the French system (CPCS, 1967, p. 25) and they may be classified in this group.

In Soil Taxonomy the concept Plaggen epipedon is defined as 'a man-made surface layer, more than 50 cm (20 inches) thick, that has been produced by long-continued manuring'. Both soils discussed have such epipedons and have to be classified as Plaggepts (SSS, 1975, p. 257). Earlier (SSS, 1960, p. 116) such soils were grouped with the Entisols as Plaggudents but this was changed in the 1967-Supplement into the present-day concept. Conry & Diamond (1971) disagree about this change. They argue that the Plaggen epipedon is not a kind of Umbric Epipedon, because 'in many cases, it has properties similar to an Ochric epipedon. A more satisfying basis is to consider the Plaggen epipedon as new parent material. . ., it is proposed that plaggensoils should be classified as Plaggents within the order Entisols'.

The Soil Survey Manual states: 'A small subscript, *an* for example, may be used for man-made or anthropic layers beneath Ap, if present, and above the undisturbed layers. Thus an anthropic humus Podzol may have horizons A1anp, A1an, A1, A2, B2, B3, and C'. (SSS, 1951, p. 182). In practice the Dutch have developed some modifications to this system. Although every part of the anthropic A1 horizon once has been ploughed, the subscript *p* is only used for the upper part of the actual plough layer. On the other hand the *p* is used for the buried A1 horizons if there is clear evidence of disturbance. In soil A2 this horizon is a mixture of material of the original A1 and A2 horizons and therefore is coded (A1+A2)pb. Furthermore in the Netherlands the first figure 1 has been dropped, Aan horizons are always A1 horizons; the horizons of the buried sola are all indicated as . .b horizons.

In most systems of soil horizon nomenclature it is not possible to use such detailed horizon notations, only the German system has a special capital letter for Aan horizons, the E (from *Esch*) with the note that if E is introduced for eluvial horizons it wil have to be changed into YA (Arbeitsgemeinschaft Bodenkunde, 1971, p. 30).

It is generally assumed that the colour of the Aan horizon is related to the kind of turves used as bedding materials in the stables. Heather turves yielded black or grey Plaggepts and grass turves produced brown Plaggepts (Pape, 1970; Mückenhausen, 1975, p. 453; Mückenhausen et al., 1977, p. 137). The analytical data support this assumption: heather sods of virgin soils have high C-N rations (cf. p. 164), whereas grass sods from brook valleys have low C-N ratios; the use as absorbing agents in the stables changed the N-content but did not obliterate the initial differences. Furthermore the heather sods have practically no clay but most soils from the brook valleys of the Pleistocene sandy district have some clay in the upper part of their sola, which is reflected in the differences between the clay content of the brown and the black plaggen soils.

Profile description

Aanp	0-25 cm	Black (7.5YR2/1) loamy fine sand; single grain; loose to very friable; abrupt, smooth boundary.
Aan2	25-75 cm	Black (7.5YR2/1.5) loamy fine sand; in situ rather massive but removed it is single grain; loose when moist, but slightly hard when dry; small fragments of pottery, bricks and many tiny fragments of charcoal; abrupt, smooth boundary.
(A1+A2)pb	75-90 cm	Mixed grey (5YR5/1), dark grey (7.5YR4/1) and very dark grey (7.5YR3/1) fine sand; reworked material, obviously the fossil plough layer, now buried under the plaggen horizon; abrupt, but somewhat irregular boundary.
B2b	90-100/115 cm	Dominantly dark brown (7.5YR3.5/4) fine sand; the colour is caused by the presence of coatings of organic matter; gradual, irregular to broken boundary.
B3b	100/115-120 cm plus	Transitional horizon to the parent material, characterized by a gradual change in colour from brown to light grey, due to a gradual thinning of the organic coatings.

The soil is non-calcareous throughout.

Analytical data

Horizon	Depth (cm)	pH-KC1	Organic fraction				Particle-size distribution of mineral fraction (% of fine earth)					
			O.m. (%)	C(%)	N(%)	C/N	< 2	2-16	16-50	50-105	105-150	> 150 µm
Aanp	0-20	3.9	6.5	2.7	0.17	15.9	6	2.5	8	16	22	46
Aan2	45-60	4.0	8.2	3.5	0.21	16.6	4.5	2.5	8	14	19	52
Cb	120-125	4.6	0.5	n.d.	n.d.	n.d.	0.3	0.4	2	17.5	25	55.7

Detailed particle-size distribution of the > 150 µm fraction of the Cb sample:

150-210	210-300	300-420	420-600	600-850	850-1200	1200-1700 µm
25	16	8	3	2.5	1	0.2

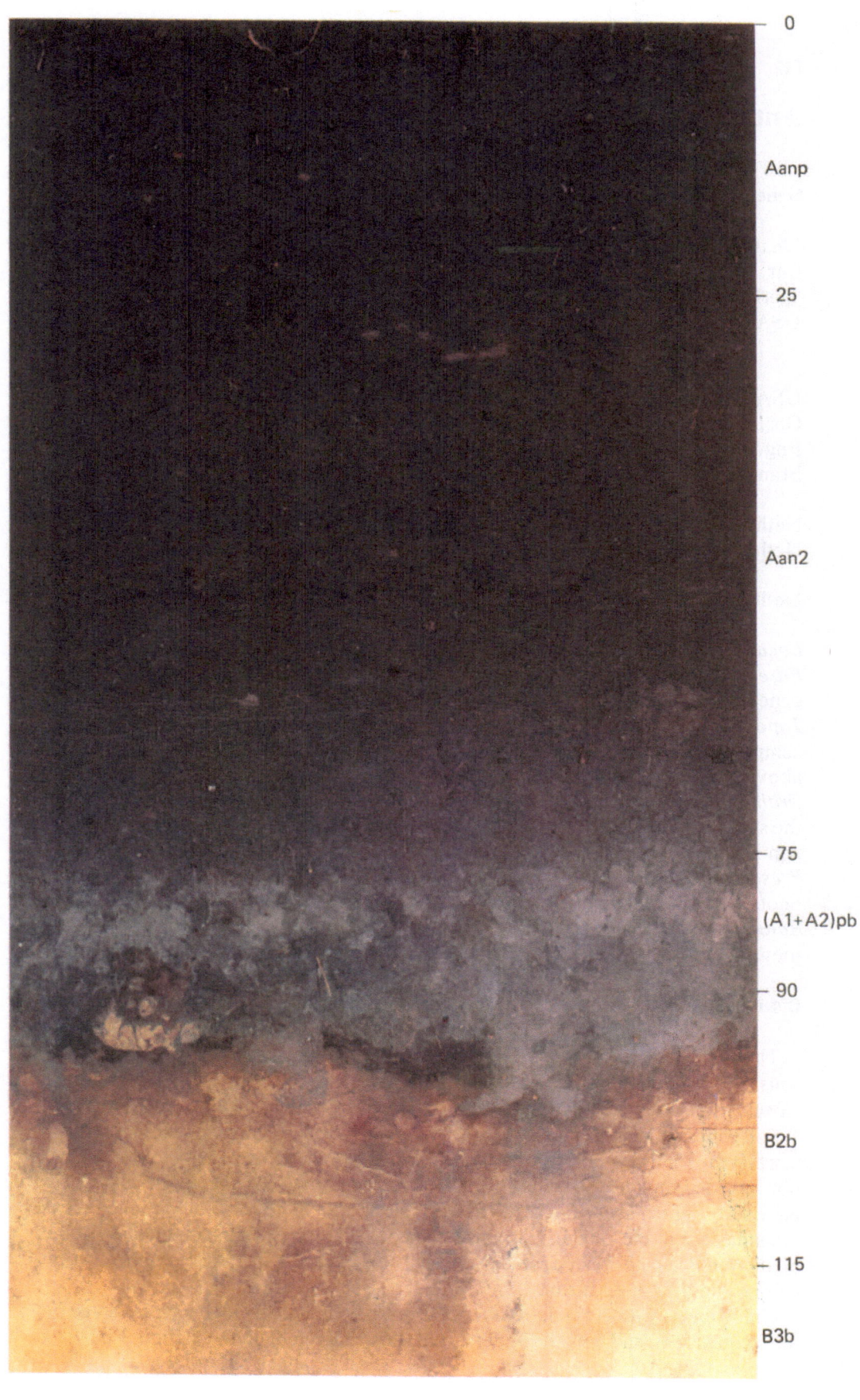

0

Aanp

25

Aan2

75

(A1+A2)pb

90

B2b

115

B3b

Anthropogenic soils: A3

General data

Classification

Eur.	Regosol (p. 3, 74 and 75)
World	Albic Arenosol (p. 34), coarse textured, gently undulating (p. 5)
USA '38	Included in the Dry Sands in the original list of classification units (p. 1001 and 1135-1137), in the 1949 modification called Regosols (p. 120)
USA '75	Siliceous, mesic Typic Udipsamment (p. 206)
Ger.	Podsol-Regosol (p. 62)
Eng. & Wales	Podzolic sand-ranker
France	Sol minéral brut d'apport éolien (p. 18) or sol d'apport éolien (p. 25)
Neth. '50	Inland dune sand (p. 22 and 23); mapping unit 47
Neth. '61	Non-calcareous sand soil, humuspoor, poor and very poor sand, not loamy; mapping unit 124
Neth. '66	'Duin' vague soil (p. 161 and 295); mapping unit Zd21

Location. Utrecht Province, about 20 km east of the town of Utrecht (see Fig. 1).
Parent material. Recent aeolian sand, locally still active dunes, overlying pleistocene aeolian sand (cover sand from Weichsel age).
Topography and elevation. Stabilized dune land with a hummocky relief. The sampling site is on a nearly level point on the crest of a low broad dune; about 10 m above sea level.
Drainage and ground water. Ground-water level is always several metres below the surface. There may be temporarily a perched water table on top of the thin iron pan in the B horizon.
Present land use. About 45 years ago the active dunes were stabilized by planting Scots pine (*Pinus sylvestris*).
Range in land use. Only a minor part of these soils are stil active dune land; mostly they are afforested with Scots pine.

Discussion

There are several reasons for including this soil in the chapter of anthropogenic soils. Most inland dunes are caused by man-induced soil erosion. These soils are closely associated with Plaggepts, geographically as well as historically (see p. 41).

Soil genesis in dune sand is practically zero because vegetation is lacking, it only starts if the sand movement is halted by afforestation. If a system of soil classification is based on soil genesis it is not clear whether active dune land is thought to have soils, which accordingly have to be classified. On many soil maps such areas are indicated in the category of 'miscellaneous land type'. Most dune lands of the Netherlands are afforested, and as this soil comes from such an area, the question will not be discussed further.

Unfortunately the 5 cm thick layer of forest litter (A0) has not been preserved on the lacquer peel, and only the upper 25 cm shows incipient podzolization. Locally a clearly developed micropodzol may be present with 2 cm bleached A2 beneath the forest litter and humus-coated sand grains down to a depth of 15-20 cm (Van der Voort, 1972, Fig. 1.4).

These phenomena justify its classification as a *Podsol-Regosol* in the second edition of the German system (formerly a *Podsol-Ranker*), and as a podzolic sand-ranker in the system of England and Wales. It must be classified as a Regosol on the Soil Map of Europe. Regosol is a term introduced in the 1949-modification of the old USA system (Thorp & Smith, 1949, p. 120). In the legend to the Soil Map of the World soils from unconsolidated materials (exclusive of recent alluvial deposits) showing weak or no soil development are either Arenosols or Regosols; because the moist colour value of the dune is high enough to be qualified as albic materials (FAO, 1974, p. 28), this soil has to be an Arenosol[1] (even the Dutch calcareous coastal dunes are no Regosols according to this criterion).

In the French system this term is reserved only for *sols bruts d'érosion sur roche meuble* (CPCS, 1967, p. 18), and the *sols minéraux bruts d'apport éolien* have no special name. Differences between classes of the *sols bruts* and the *sols peu évolués* are very slight: 'only traces of organic matter' (CPCS, 1967, p. 17), and: 'more than traces of organic matter' (CPCS, 1967, p. 21). This soil possibly already shows enough evidence of soil genesis to be named *peu évolué*.

Its classification in Soil Taxonomy yields no difficulties: lack of diagnostic horizons makes it an Entisol, its sandy texture a Psamment, and as the soil moisture regime is clearly Udic in this Atlantic climate (Table 2), this makes it an Udipsamment. Because lamellae, mottles, solid rock and an albic horizon (SSS, 1975, p. 206) are all absent, it belongs to the typic subgroup of the Udipsamments.

The depth of the buried soil and the content of organic matter in the dune sand are very important for the growth of the trees that were planted on many inland-dune areas in the Netherlands 40-60 years ago. The subdivision on detailed soil maps is based on these two aspects (Schelling, 1955).

The first 25 cm show no clear horizon differentiation, so this horizon has only been given a number (SSS, 1951, p. 174). The stratified dune sand is coded C, and the horizons of the overblown podzol have the letter suffix b. This buried soil is slightly truncated: the A0 and the upper part of the A2 are lacking (especially conspicuous in the left part of the picture). Between the A2b and the B2b there is a thin iron pan similar to that in soil P3 (p. 162-165), which has to be coded a Bms (FAO, 1974, p. 22).

Analytical data of inland dune sand are straightforward. A sample from the C horizon shows some of its typical characteristics: no lime, low pH and the particle-size distribution indicates it originated from Pleistocene aeolian sand (cf. p. 156). The organic-matter content depends on the number and thickness of the humic layers, which are few and thin in the major part of this C horizon. If they are many and thick as in the lower 10 cm of this C horizon, a mixed sample may have up to 3-5% organic matter.

---

1. The definitions of the Arenosols and the Regosols have to be changed in such a way that these dune sands key out in the Regosols (Dudal, personal communication, 1978).

Profile description

1	0-25 cm	Variegated grey (10YR5/1), light brownish grey (10YR6/2) and brown (10YR5/3) fine sand; in the brown patches sand grains with thin coatings of organic matter, in the grey patches the sand is bleached, both indicating the very beginning of podzolisation; original stratification disturbed; loose; gradual, wavy boundary.
C	25-98 cm	Greyish brown (10YR5/2), loose fine sand; stratified, stratification recognisable as small differences in grain size, silt content and organic matter content; just above the buried podzol there are thin layers with more organic matter; very abrupt, smooth boundary.
A2b	98-106 cm	Grey (5YR5.5/1) loose fine sand, strongly bleached sand grains stained with some organic matter; at the very abrupt smooth boundary a very thin discontinuous iron pan.
B2b	106-120 cm plus	Dark reddish brown (5YR5/3) fine sand, with very thin coatings of iron and of organic matter; firm; variegated with light grey (7.5YR7/1) loose, noncoated fine sand.

The Udipsamment, as well as the buried Spodosol, are non-calcareous.

Analytical data

Hori-zon	Depth (cm)	pH-KCl	CaCO₃ (%)	O.m. (%)	Particle-size distribution of mineral fraction (% of fine earth)							
					<50	50-75	75-110	110-150	150-210	210-300	300-420	> 420 µm
C	40-50	4.2	0	0.6	1.4	1.5	6	26	31	21	11	2

0

1

—25

C

—98

A2b

—106

B2b

Anthropogenic soils: A4

General data

Classification

Eur.	Anthropogenic modification of an alluvial soil; not mentioned as such in the description of the alluvial soils (p. 4), and on the map they are included in this association (p. 75-77)
World	Not mentioned in the list of soil units
USA '38	Not mentioned, neither in the 1938 original list of classification units, nor in the 1949 modification
USA '75	Fine-loamy, mixed, mesic Plaggept (p. 257)
Ger.	Hortisol (p. 139)
Eng. & Wales	Man-made humus soil (no appropriate category at subgroup level)
France	Sol d'apport anthropique (p. 25)
Neth. '50	Man-made soil; mapping unit 54 (p. 119)
Neth. '61	Calcareous young sea clay soil, deeply humose, with a heavier subsoil, clayey sand to light clay; mapping unit 18
Neth. '66	'Tuin' earth soil (p. 146 and 192); mapping unit EK19

Location. Zuid-Holland Province, a few km south of The Hague, in a polder embanked before 1300 A.D. (see Fig. 1).

Parent material. Recent marine sediment, with superficial admixture of dune sand.

Topography and elevation. Level; about sea level.

Drainage and ground water. The field is tile drained at an interval of 10-12 m, further drained by ditches spaced 50-75 m apart into a relict tidal creek (see Fig. 26). From this creek the water is pumped into the *Nieuwe Waterweg* (8 km to the south), the shipping lane from Rotterdam to the North Sea. The ground-water level fluctuates between 60 cm and 1.25 m depth.

Present land use. The sampling site (Fig. 26) is on a field with hothouses and Dutch frames, open space was lettuce; horticultural land since about 1800 A.D.

Range in land use. Exclusively for horticulture in glasshouses (the area is called the Westland Glass District). Important crops are lettuce, tomatoes and cucumbers, and flowers like chrysanthemums, freesias and carnations. Twenty years ago grapes were important, but nowadays few are grown.

Discussion

This kind of soil is found exclusively in Westland, the horticultural district between Rotterdam, Hook of Holland and The Hague (Fig. 26). Compared with the classic plaggen soils it is not extensive in the Netherlands (220 000 ha and 5400 ha, respectively). Like soil A1 (p. 134-137) and soil A2 (p. 138-141) its surface level has been gradually raised by man. However, it originated as a calcareous soil in Holocene marine silt loams (such as soil M6, p. 82-85), not as a podzol in Pleistocene aeolian sand. Its surface was raised by applying dune sand mixed with mud dredged from the ditches, not by heather or grass sods used in the stables, and it has been under continuous use as horticultural land.

In his review of plaggen soils Conry (1974) discusses not only the classic plaggen soils of the Pleistocene sandy districts of northwestern Europe, but also other soils gradually raised by man. One of these he studied himself, viz. the so-called sanded soils, which '. . . did not originate from an admixture of plaggen material and farmyard manure but from the prolonged application of sea-sand either alone or in conjunction with farmyard manure, sea-weed or even peat mould' (Conry, 1969). These soils '. . . occur only in coastal districts of Ireland, contiguous to good sources of calcareous sea sand. They generally occur within a mile of the sea coast.' (Conry 1971). These Irish soils have more in common with the Dutch soils discussed in this section than with the classic plaggen soils.

All these soils satisfy the definition of the Plaggepts of Soil Taxonomy, but the German system separates the soils produced '. . . by decades – respectively ages – long intensive horticulture, . . .' (Mückenhausen et al., 1977, p. 139) from the *Plaggenesch* on the level of the *Typ* (comparable to the level of the great group in Soil Taxonomy). These *Hortisols* include soils with A1 horizons deepened by intensive deep cultivation and biological activity, whereas Conry & Diamond (1971) want to exclude such soils in their proposal for the subdivision of the Plaggepts.

In the French system there is the problem, discussed on p. 138, of the relevance of soils such as these to the *groupe des sols anthropiques*.

Because there is no subgroup for loamy man-made humus soils, it is not possible to classify this soil in the system of England and Wales beyond the group level, unless these soils are placed in the categories sandy or earthy man-made humus soils, neither of which are the correct places for this soil.

Elucidation of the use of the letter subscripts an, p and b in the horizon nomenclature of raised soils is given in the discussion about plaggen soils (p. 139). When describing the Irish sanded soils Conry (1971) uses the following codes: Ap11-Ap12-Ap2b-Cb.

From the particle-size distribution, it is clear that dune sand was used; the buried soil has 23% clay and 9% coarser than 150 μm, quite normal for recent marine sediments (cf. the analytical data on p. 76, 84 and 112). The Aanp has been 'diluted' with dune sand to such an extent that the original clay content has decreased from 23 to 6%, and the sand separate has increased from 9 to 67%.

Although the buried soil is calcareous, as was the applied dune sand, the lime content decreases upwards. This phenomenon is most probably due to the prolonged and exclusive use of sulphate of ammonia, an acidifying nitrogen fertilizer.

Other analytical data are normal, including a low C-N ratio, high percentage of exchangeable calcium and an exchange capacity of 60-70 meq per 100 g clay.

Profile description

Aanp	0-20 cm	Dark grey brown (2.5Y4/2) loamy fine sand; very weak structure not readily recognizable, nearly massive; when disturbed very friable and soft; abrupt, smooth boundary.
Aan2	20-70 cm	Dark grey brown (2.5Y4/2) fine sandy loam, the whole horizon is characterised by a gradual decrease with depth of the dune-sand separate (150-210 μm); small pieces of brick; grade of structure is higher than in overlying horizon: weak, fine subangular blocky, locally massive; friable, soft; abrupt, smooth boundary.
Apbg	70-90 cm	Very dark grey (2.5Y3.5/1) loam; slightly mottled; weak, coarse prisms, breaking to very weak fine subangular blocky structure; friable to firm, slightly hard; most probable the original topsoil; abrupt, smooth boundary.
C2g	90-120 cm plus	Grey (2.5Y5/1) silt loam; few to common fine distinct brown mottles; fine shell grit; moderate ompound coarse prisms breaking to fine subangular blocky structure; friable, slightly hard.

Analytical data

Horizon	Depth (cm)	Organic fraction				Particle-size distribution of mineral fraction (% of fine earth)					
		O.m. (%)	C(%)	N(%)	C/N	< 2	2-16	16-50	50-105	105-150	> 150 μm
Aanp	0-20	1.4	0.8	0.08	10	6	4.5	10	4	8	67
Aan2	40-60	1.2	n.d.	n.d.	n.d.	16	9.5	15.5	7.5	9	42
Apbg	80-90	1.3	n.d.	n.d.	n.d.	23	14	32	14	8	9
C2bg	110-120	0.5	n.d.	n.d.	n.d.	23	15.5	35	14	8	6.5

Horizon	CaCO₃ (%)	pH-KC1	Extractable cations (% of sum)					C.E.C. (meq./100 g)	Ca/Mg	K-fix. (%)
			Na	K	Mg	Ca	H			
Aanp	0	4.7	n.d.	n.d.	n.d.	n.d.	n.d.	n.d.	n.d.	n.d.
Aan2	4.0	4	3.1	1.6	7.8	85.3	2.2	12.9	11	n.d.
Apbg	7.1	7.5	n.d.	n.d.	n.d.	n.d.	n.d.	n.d.	n.d.	n.d.
C2bg	11.8	7.6	1.9	1.2	0.6	93.8	2.5	16.2	152	24

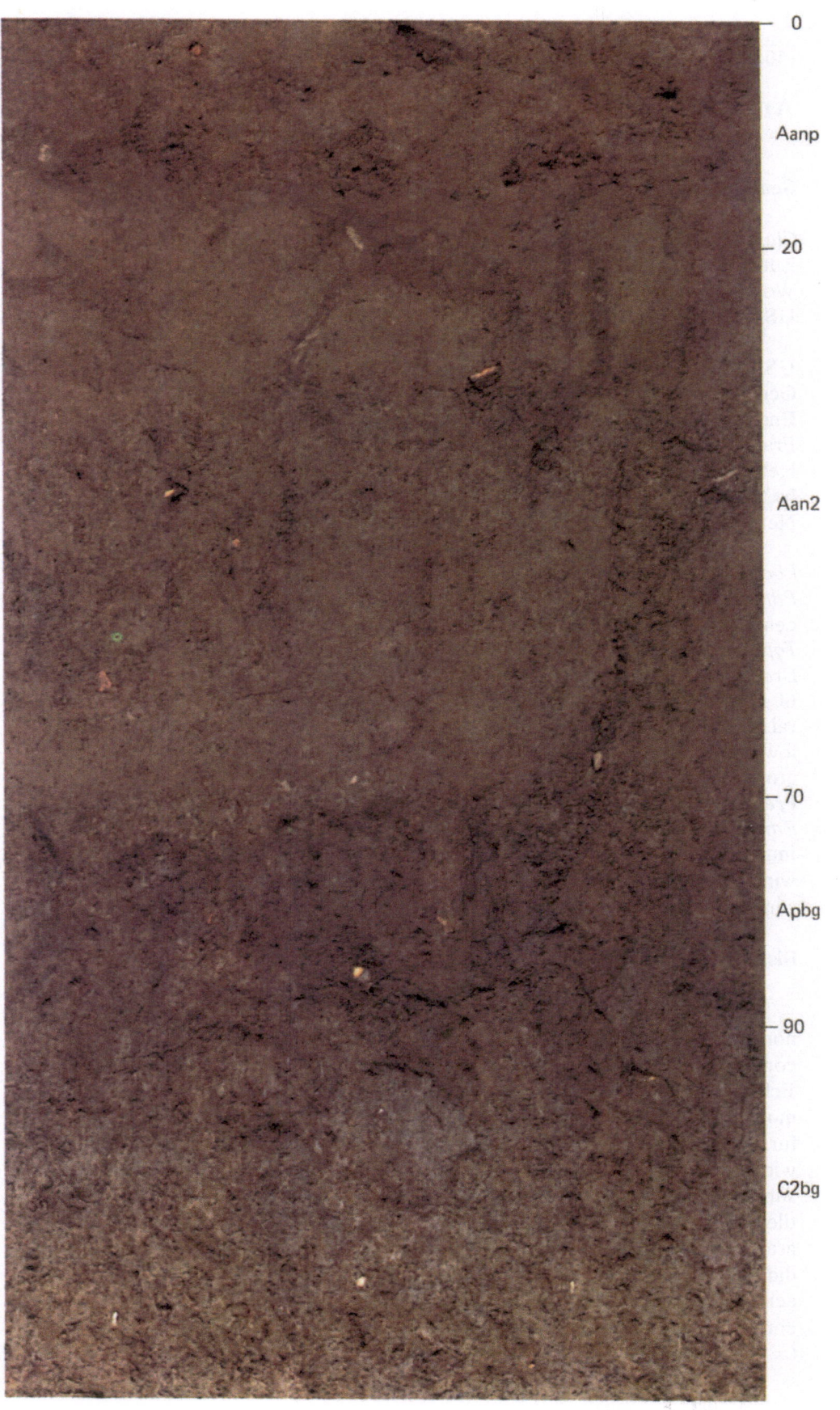

0

Aanp

— 20

Aan2

— 70

Apbg

— 90

C2bg

Anthropogenic soils: A5

General data

Classification

Eur.	No suitable category
World	No suitable category
USA '38	No suitable category, either in the original list of classification units (1938), or in the 1949 modification
USA '75	Coarse-silty, mixed, mesic Plaggept (p. 257)
Ger.	Hortisol (p. 139)
Eng. & Wales	Man-made humus soil, no appropriate category at subgroup level
France	Sol d'apport anthropique (p. 25)
Neth. '50	Not recognized at that time, included in mapping unit 13
Neth. '61	Not recognized at that time, included in mapping unit 19
Neth. '66	'Tuin' earth soil (p. 146 and 192); mapping unit bMn15C

Location. Friesland Province (see Fig. 1).
Parent material. Recent marine sediment, probably embanked since the 12th century.
Topography and elevation. Level; about 1 m above sea level.
Drainage and ground water. Not tile drained but drained by ditches spaced about 60 m apart, discharging into the main drainage channel from which the water is raised by pumping into the Wadden Sea. This is an extensive foreshore, dry during low tide except for the tidal creeks, which stretches out to the Frisian Islands. The ground-water level fluctuates considerably between 50 cm and 1.80 m.
Present land use. Grassland, used both for pasture and meadow.
Range in land use. These soils are predominantly used as arable land, a very important crop is seed potatoes, other crops are sugar beet and small grains, mainly winter wheat; a small area is used for horticulture with a variety of vegetables, some under glass. Bulb-growing (tulips) is beginning in this district.

Discussion

This soil was included in this inventory of Dutch soils to illustrate a soil from the north of the country, developed from old marine sediments characterized by a high content of very fine sand. Such soils need to be tile-drained at a very close interval. Before introduction of tile drainage (in the last quarter of the nineteenth century) most of the fields in the marine district of the Netherlands had a kind of ridge-and-furrow drainage system, every field having 5 to 10 of these ridges, each 10-20 m wide with shallow narrow open-field drains between, draining into the ditches surrounding the fields. In grassland areas this pattern still exists, but in arable areas tiles have been laid below the furrows and the field drains have been levelled. The arable land in the northern marine district is an exception because there the fields did not have open-drain systems, but the *whole field was made convex*. This was achieved by raising the middle of the field with soil from the margins and the corners (Fig. 38), thus creating fields with a domed surface, called *kruinig* in Dutch (= cresty).

These fields are comparable with the *champs bombés* northeast of Ghent, described by Snacken (1971). The soil of plate A5 was sampled on the crest of such a domed field in the northern marine district in the Netherlands.

Compared with other raised soils (soil A1, A2 and A4) this man-made soil has quite a different genesis and geography, but the discussion about its classification is similar.

The land use is grassland, and as no former Ap is recognizable, the upper horizon (the sod) is coded Aan. The organic-matter content is low in the plough layers of marine soils; their Munsell values are mostly 4-4.5 (cf. p. 76 and 84), so the buried Ap is hardly discernible; and must occur in the lower 20-25 cm of the Aan2 horizon.

In contrast to the low organic-matter content of plough layers in these soils, this grassland has a high content in the turf (9.5%). The difference in CEC between the upper two samples is entirely due to the difference in organic matter content, the clay contents of both samples being the same. Values for CEC of both samples are in accordance with the empirical formula: CEC=0.6 to 0.8 (%clay + 3 times % o.m.).

The lime content of the Aan horizons is lower than in the subsoil horizons. This may be explained by the removal of the shallowly decalcified soil from the sites of the ditches and the corners of the fields to the centre revealing unleached calcareous material whereas the Aan horizons in the centre of the field are lacking in lime.

Compared with the other samples of marine soils of approximately the same clay content, these samples have a much finer sand fraction, only slightly coarser than loess.

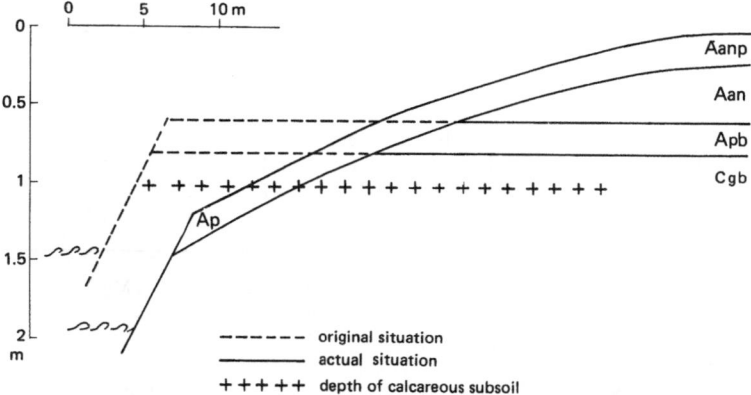

Fig. 38. Sketch-section through a convex field from ditch (left) to crest (right). Adjacent to the ditch the surface has been lowered and the new Ap has developed in the former subsoil. On the crest the original Ap is buried beneath a kind of Plaggen epipedon.

Profile description

Aan1	0-12 cm	Very dark greyish brown (2.5Y3.5/2) loam; weak, very fine to fine subangular blocky; very friable, soft; clear, smooth boundary.
Aan2	12-86 cm	Dark greyish brown (2.5Y4.5/2) loam; fine shell grit, many worm holes partly filled in with dark material, small pieces of bricks, at 45 cm depth greyish silty material; weak, very fine subangular blocky; very friable, soft (no buried A-horizon recognisable); clear, smooth boundary.
C2g	86-120 cm plus	Grey (2.5Y6/1) silt loam grading via loam to very fine sandy loam; stratified, fine shell grit; mottles both horizontal ('following the stratification') and vertical alongside root channels; no pronounced structure, a tendency towards fine subangular blocky; friable, soft.

Analytical data

Hori-zon	Depth (cm)	Organic fraction				Particle-size distribution of mineral fraction (% of fine earth)				
		O.m. (%)	C(%)	N(%)	C/N	< 2	2-16	16-50	50-105	> 105 μm
Aan1	0-10	9.5	4.3	0.46	9.3	14	6	32	43	5
Aan2	20-45	1.9	n.d.	n.d.	n.d.	14	8	34	39	5
C21g	86-100	0.7	n.d.	n.d.	n.d.	19	11	47	22	1
C22g	110-120	0.4	n.d.	n.d.	n.d.	10	4.5	26	54	4.5

Hori-zon	$CaCO_3$ (%)	pH-KCl	Extractable cations (% of sum)					C.E.C. (meq./ 100 g)	Ca/Mg	K-fix. (%)
			Na	K	Mg	Ca	H			
Aan1	0.7	6.5	0.7	2.8	8.7	79.5	8.3	28.8	9.0	n.d.
Aan2	2.4	6.8	2.1	7.6	7.6	79.2	3.5	14.4	10.4	n.d.
C21g	9.6	7.3	2.5	1.3	7.5	87.0	1.3	16.2	11.6	n.d.
C22g	8.3	7.4	n.d.	n.d.	n.d.	n.d.	n.d.	n.d.	n.d.	26

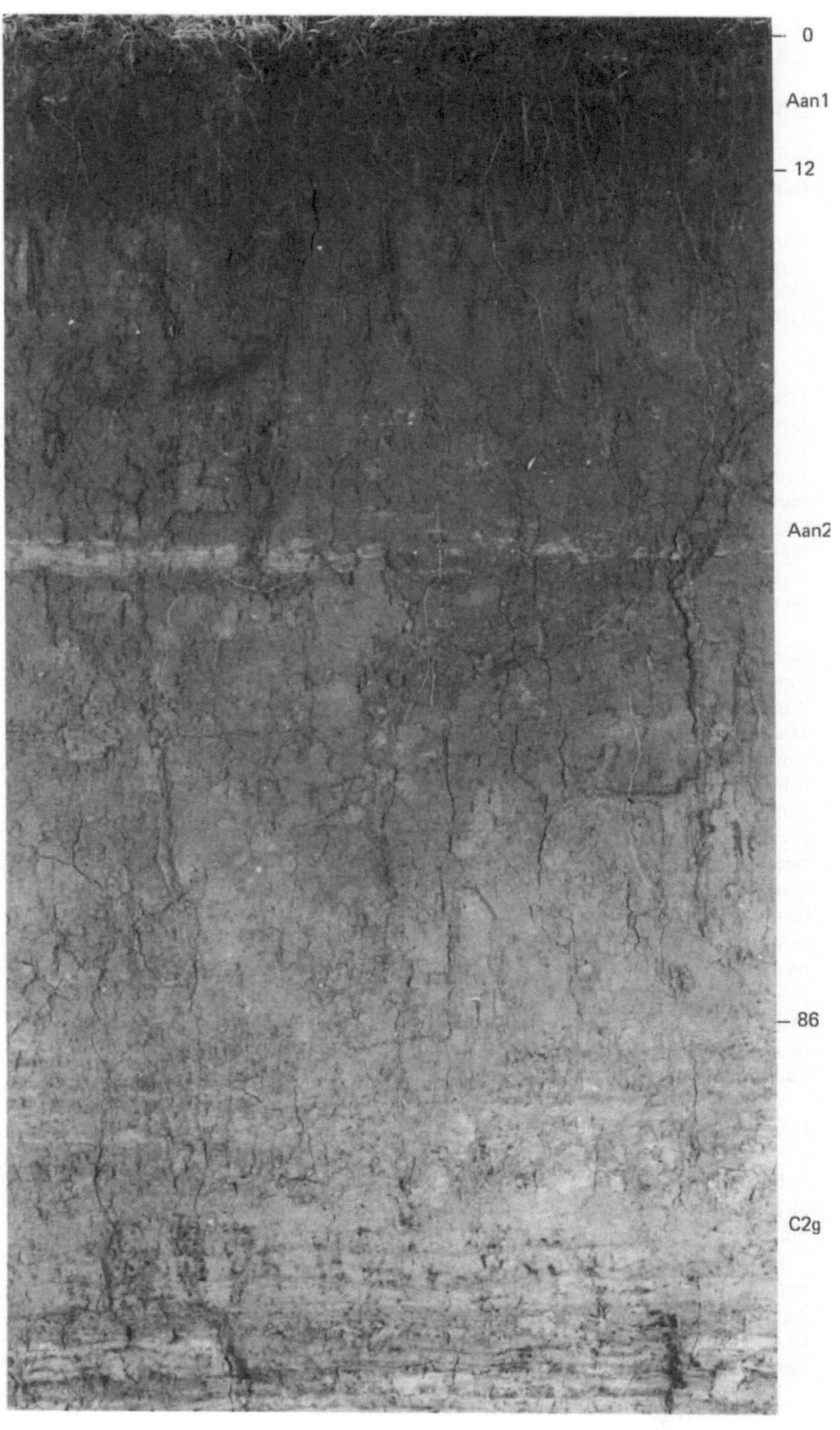

0

Aan1

12

Aan2

86

C2g

Soils of the Pleistocene sandy district: P1

General data

Classification

Eur.	Podzolized soil (p. 5 and 6)
World	Gleyic Podzol (p. 39), coarse textured, level to undulating (p. 5)
USA '38	Not a Podzol, because it is hydromorphic, but also not a Ground-Water Podzol, because there is no 'whitish-gray leached layer up to 2 or 3 feet thick' (p. 1000)
USA '75	Sandy, siliceous, mesic Typic Haplaquod (p. 336 and 337)
Ger.	Gley-Podsol (p. 111)
Eng. & Wales	Typical (humic) gley-podzol
France	Podzol humique à gley (p. 54)
Neth. '50	Moist sandy soils; mapping unit 39, called wet or low heatherprofile (p. 26-29)
Neth. '61	Non-calcareous, low humuspodzol, very poor, slightly loamy sand; mapping unit 101
Neth. '66	'Veld' podzol soil (p. 130, 131 and 189); mapping unit Hn21

Location. Utrecht Province, about 20 km east of the town of Utrecht (see Fig. 1).
Parent material. Cover sand, i.e. an aeolian sand from the Weichsel age.
Topography and elevation. Nearly level, about 7 m above sea level.
Drainage and ground water. Drained by a few shallow ditches and furrows, no runoff, most of the superfluous precipitation sinks into the soil. The area is drained off by a brook in the immediate surroundings, which is lying about 2 m lower in the same geological formation. The water table fluctuates between 30 cm and 1 m.
Present land use. Grassland used as pasture which was reclaimed from marshy heathland with a dominant vegetation of cross-leaved heath (*Erica tetralix*) and ling (*Calluna vulgaris*) in about 1910.
Range in land use. Most of these soils were reclaimed between 1850 and 1950, very few areas still remain marshy heathlands. Used both as grassland (pasture as well as meadow) and arable land; the acreage of grassland has increased in the last ten years, mainly long-term leys. Main arable crops were until recently rye, potatoes, little sugar beet and wheat. Silage-maize has become a more important crop in the last few years. There is some afforestation on these soils, mainly Scots pine with larch and Douglas fir.

Discussion

Because this soil has a 'B-horizon showing. . . humus accumulation' (Dudal et al., 1966, p. 5) it is a podzolized soil on the Soil Map of Europe. The soil cannot be placed in the old USA system, for the reasons given above.

Part of item 2 of the definitions of both the spodic B horizon in the FAO system and the spodic horizon of Soil Taxonomy (which are nearly identical) reads: 'A sandy or coarse loamy texture . . . with sand grains covered with cracked coatings' (FAO, 1974, p. 26). This is enough to identify the soil as a Podzol (FAO) or a Spodosol (Soil Taxonomy).

The brownish colours of the B horizon are caused solely by humus coatings. Iron coatings are lacking; this is clearly demonstrated by the whitish colour of the inset in the colour plate, a photo of an undisturbed ignited sample (Van der Voort, 1972). This property is diagnostic of a Gleyic Podzol (FAO) or an Aquod (Soil Taxonomy). On the two lower levels in the American system the soil is 'double normal', making it a Typic Haplaquod.

At the highest level of the German system (*Abteilung*) the soil is a *Terrestrischer Boden*; in this class 'are grouped the soils originating from soil forming processes beyond the reach of the ground water' (Mückenhausen et al., 1977, p. 51). At the level of the *Subtyp* (comparable to subgroup level in most other systems) it is a *Gley-Podsol,* thus being a soil 'which is a podzol in the upper part of the solum and a gley soil in the lower part' (Mückenhausen et al., 1977, p. 111). As there is no resemblance between the subsoil of this soil and that of a *Gley* such as soil P6 (p. 177) these statements are questionable, particularly the latter half of the second quotation. For, a characteristic of the lower subsoil of a *Gley* is the removal of iron from the individual particles of sand by the process of reduction, thus exposing the bluish, nearly neutral colours of the clean quartz grains in the Gr horizon, and in the overlying Go horizon the iron segregates into brown 'rusty' mottles. Such phenomena are clearly lacking in this podzol soil.

Realizing that the typical non-hydromorphic podzol (soil P3, p. 165) has lost its iron only in the upper 23 cm, and has thinner organic coatings at shallower depth, it must be concluded that hydromorphic podzols are more strongly podzolized than the non-hydromorphic podzols.

In the French system as well that of England and Wales there are specific subgroups for these kind of soils, but in both systems it is stated: 'Below the Bh, there is a gleyed Go or Gr' (CPCS, 1967, p. 54), and: 'gleyed horizon directly below the podzolic B or at less than 50 cm' (Avery, 1973, p. 327). However, there are no G horizons in the hydromorphic podzol soils in Les Landes of southwest France, nor is there a gleyed horizon directly below the B horizon of the soils of the Crannymoor series in the Lancashire Coastal Plain of England (Ragg & Clayden, 1973, p. 116).

The A horizon has strongly bleached sand grains; none of the discussed systems of soil horizon nomenclature can indicate this. Possibly Eh or EA, meaning a humose E horizon or a bleached A horizon, could be used. Loss of iron is one of the features of an A2 (Ae or E) horizon. In this respect all soil horizons of this soil could be coded E. Most probably this extreme situation has been caused by a combination of both the gleying and the podzolizing processes. There is no symbol of 'gleying' without reduction colours of mottling either.

The low iron content of the AB and B2 samples confirm the visual picture of the ignition strip on the photograph. The upper part of it (the lower few cm of the Ap) is slightly pinkish, and has a slightly higher iron content.

Profile description

A11	0-8 cm	Very dark grey (7.5YR3/1) fine sand, the sand grains are strongly bleached and turn white on ignition, but the bleaching is concealed by the high organic matter content; single grain to moderate weak fine subangular blocky; very friable, soft; abrupt, smooth boundary.
Ap	8-20 cm	Very dark grey (7.5YR3/1) fine sand, locally lumps of black (5YR2/1) mucky fine sand (most probable remnants of the original heather sod); less organic matter than in the A11, the bleached character is more conspicuous; single grain to very weak fine subangular blocky; very friable, loose; abrupt, somewhat irregular boundary (spade or plough marks).
AB	20-30 cm	Reddish grey (5YR5/2) loamy fine sand; a transitional horizon with in the upper part still bleached sand grains, in the lower part the sand grains are coated with organic matter.
B2	30-55 cm	Dark brown (7.5YR3/4) fine sand; massive, but easily crushed to single grain; loose; all sand grains coated with amorphous humus and slightly cemented on the points of contact by bridges of humus, no iron coatings; many vertical streaks filled in with organic matter, most probable a fossil root system; gradual, smoot boundary.
B3	55-120 cm plus	Dark yellowish brown (10YR4.5/4) fine sand, characterized by a gradual thinning of the organic coatings, no iron coatings; in situ massive but easily crushed to single grain.

The soil is non-calcareous throughout.

Analytical data

Hori-zon	Depth (cm)	Fe$_2$O$_3$ (%)	pH-KC1	Organic fraction				Particle-size distribution of mineral fraction (% of fine earth)				
				O.m. (%)	C(%)	N(%)	C/N	< 2	2-50	50-105	105-150	> 150 μm
A11	0-8	0.15	4.6	11.7	4.7	0.38	12.4	4	6.5	4.5	16	69
Ap	8-20	0.26	4.3	9.3	4.2	0.24	17.5	1.5	7.5	4.5	13	73.5
AB	20-30	0.03	3.7	4.7	2.1	0.10	21.0	3	12.5	10.5	19	55
B2	30-55	0.04	3.9	2.0	1.1	0.04	27.5	0	0.5	5	18	76.5
B3	60-80	0.07	4.2	0.8	n.d.	n.d.	n.d.	0	0.5	2.5	15	82

Detailed particle-size distribution of the > 150 μm fraction of the B3 sample:

150-210	210-300	300-420	420-600	> 600 μm
28	36	14	4	0.7

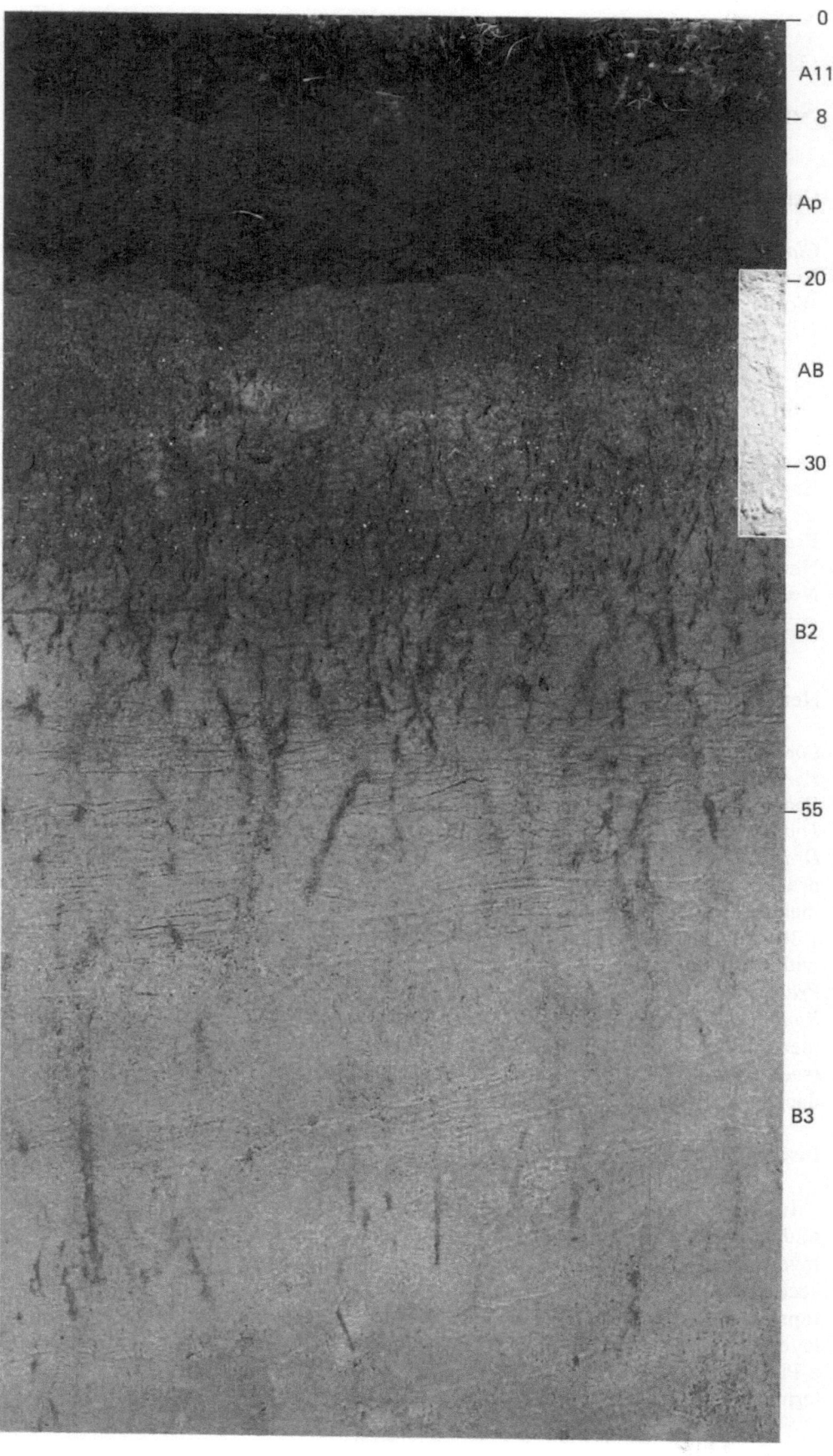

Soils of the Pleistocene sandy district: P2

General data

Classification

Eur.	Podzolized soil (p. 5 and 6)
World	Not mentioned specifically, is included in the Gleyic Podzols
USA '38	Not a Podzol, because it is hydromorphic, but also not a Ground-Water Podzol, because there is no 'whitish gray leached layer up to 2 or 3 feet thick' (p. 1000)
USA '75	Sandy over loamy, siliceous, mesic Plaggeptic (?) Haplaquod (p. 336 and 337)
Ger.	Plaggenesch-Podsol (p. 111)
Eng. & Wales	Not mentioned specifically, it is an intergrade between a humus podzol and a sandy man-made humus soil
France	Not mentioned specifically, included in the podzol humique à gley
Neth. '50	Not mentioned specifically
Neth. '61	Non-calcareous, medium high podzol, very poor slightly loamy fine sand with a humose topsoil of old reclamations of 30-50 cm thickness, overlying boulder clay which begins within 125 cm, mapping unit 190js
Neth. '66	'Laar' podzol soil (p. 128 and 189), mapping unit cHn23x

Location. Friesland Province (see Fig. 1).

Parent material. Moderately shallow cover sand Weichsel age over glacial till from Saale age.

Topography and elevation. Level; about 4 m above sea level.

Drainage and ground water. Drained by ditches spaced 40-60 m apart. The ditches drain by gravity into a semi-natural stream, which discharges into Lake Yssel by means of a pumping station. During sampling no ground water was observed within 1.30 m; during wet periods there is a perched water table above the firm till, but not within 30 cm depth.

Present land use. Grassland.

Range in land use. Dominantly grassland with alternating use as pasture and meadow, both for hay and silage-grass. The main arable crops include potatoes (seed potatoes and starch potatoes), sugar beet, oats, wheat and rye. In the last decade silage-maize has become an important crop.

Discussion

In the Pleistocene district in the north of the country the cover sands are underlain by glacial till (Fig. 2). One of the reasons for including soil P2 in this inventory of Dutch soils is to illustrate this particular geological condition. The second reason is that this soil has been raised by plaggen manure, but the dark topsoil is too thin (38 cm) to qualify as a Plaggen epipedon ('a man-made surface layer, more dan 50 cm thick', SSS, 1975, p. 18), thus forming an intergrade between a Plaggept (like soil A2, p. 141) and a podzol that has been reclaimed after the termination of the use of plaggen manure (like soil P1, p. 157). The third reason is

that this soil has a mottled subsoil (the Dg), which is quite uncommon in hydromorphic podzol soils in the Netherlands.

The mottled horizon in the underlying glacial till, unlike the subsoil of soil P1, has
not lost its iron. Only Aquods developed in deep cover sands show the typical
properties of having 'no coatings of iron oxides on the individual grains of silt and
sand in or immediately below the spodic horizon' (SSS, 1975, p. 333). As soon as
there is a finer textured subsoil (like loess, tertiary clay or this glacial till) or a
subsoil with more weatherable minerals (like Pleistocene river sands), the classic
hydromorphic properties are present: 'fine or medium mottles of iron and manga-
nese in the materials immediately below the spodic horizon' (SSS, 1975, p. 333).

The intergrade character of this soil finds expression in the German system
where there is a definite place for it: a 'podzol with a plaggen layer up to 40 cm
thickness' (Mückenhausen et al., 1977, p. 111).

In Soil Taxonomy there exists a Plaggeptic Haplohumod, having 'a surface
horizon > 30 cm thick that meets all the requirements for a plaggen epipedon
except thickness' (SSS, 1975, p. 342). Spodosols with such a surface horizon are
seldom Humods, but mostly Aquods. So it is proposed that this soil be called a
Plaggeptic (?) Haplaquod (the question mark indicates that such a subgroup has not
been defined in Soil Taxonomy, but is being proposed by the author).

In the other systems there are no special classes for this intergrade with a
medium thick A1 horizon and the soil has to be classified appropriately.

As in soil A1 (p. 137) and A2 (p. 141) the partly man-made topsoil is indicated by
Aan, meaning an anthropogenic A1 horizon. This symbol which was adopted from
the Soil Survey Manual (SSS, 1951, p. 182), but which has been omitted in Soil
Taxonomy and is also not present in the other discussed systems. Strictly speaking, every Aan horizon is an Ap, because this horizon has been produced by using
the earth-containing plaggen manure (see also p. 134) on arable land. This is
apparent on the plate from the plough or spade marks in the transition zone
between the Aan2 and the AB horizons.

The solum has been developed exclusively in the cover sand and rests directly on
the glacial till (Dg or IICg horizon).

The soil has been used for grassland for some decades and the soil-forming
processes under grassland are reflected by the difference in organic-matter content
between the sod (Aan1) and the second horizon (Aan2). The C-N ratio increases a
little with depth, an uncommon phenomenon in grassland soils; in this case it is a
relic from the original podzol properties: the C-N ratio of organic matter derived
from the original vegetation (heath) is wider than from a grass vegetation.

Profile description

Aan1	0-5 cm	Very dark grey (10YR3/1) fine sandy loam; many bleached sand grains; very weak, very fine to fine subangular blocky; friable, soft; abrupt, smooth boundary.

Aan1   0-5 cm   Very dark grey (10YR3/1) fine sandy loam; many bleached sand grains; very weak, very fine to fine subangular blocky; friable, soft; abrupt, smooth boundary.

Aan2   5-38 cm   Very dark grey (10YR3.5/1) loamy fine sand; many bleached sand grains, less organic matter; very weak, very fine subangular blocky to single grain; very friable, loose to soft; abrupt, somewhat irregular boundary (spade marks).

AB   3-48 cm   Grey (7.5YR5/1) fine sand; intermediate horizon: bleached sand grains in the upper part, in the lower part the sand grains are coated; clear, smooth boundary.

B2   48-55 cm   Dark reddish brown (5YR3/4) fine sand; rather massive but not cemented, easily crushed to single grain; all sand grains strongly coated with amorphous humus, no iron coatings; clear, smooth boundary.

B3   55-75 cm   Reddish brown (5YR4.5/4) fine sand, thinner coatings of humus than in the overlying horizon; abrupt, wavy boundary to the glacial till.

Dg   75-120 cm   Grey (10YR5.5/1) sandy loam; massive; firm, hard to very hard; common, prominent, medium brown mottles (glacial till).

The whole soil, both solum and substratum, is non-calcareous.

Analytical data

Hori-zon	Depth (cm)	pH-KCl	Organic fraction				Particle-size distribution of mineral fraction (% of fine earth)					
			O.m. (%)	C(%)	N(%)	C/N	< 2	2-16	16-50	50-105	105-150	> 150 µm
Aan1	0-5	4.8	12.1	4.8	0.42	11.4	5	3	22	16	13	42
Aan2	10-25	4.4	9.0	4.1	0.28	14.6	4	3	15	17	19	43
B2	50-55	4.2	2.5	n.d.	n.d.	n.d.	2	1	4.5	16.5	21	55
Dg	100-110	4.2	0.4	n.d.	n.d.	n.d.	18	7.5	7.5	15	14	38

Detailed particle-size distribution of the > 150 µm fraction of the B2 and Dg samples:

	150-210	210-300	300-420	420-600	600-850	850-1200	1200-1700 µm
B2	25	18	8	2	1	0.4	0.2
Dg	14	12	7	2.5	1.5	1	0.3

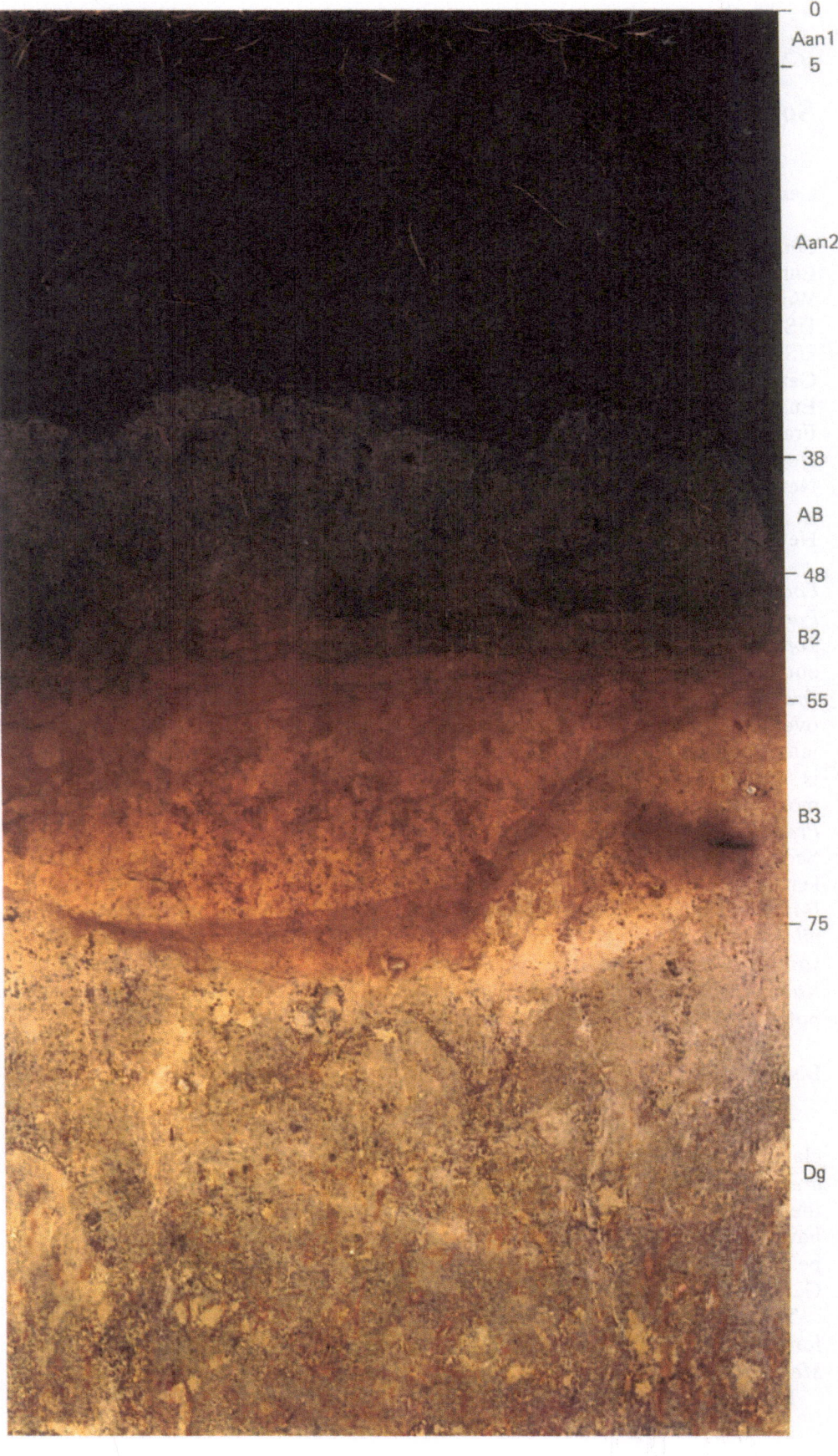

0

Aan1

5

Aan2

38

AB

48

B2

55

B3

75

Dg

Soils of the Pleistocene sandy district: P3

General data

Classification

Eur.	Podzolized soil (p. 5 and 6)
World	Humic Podzol (p. 39), coarse textured, level to undulating (p. 5)
USA 38	Podzol soil (p. 997 and 1020)
USA '75	Sandy, siliceous, mesic Typic Haplohumod (p. 341 and 342)
Ger.	Eisenhumuspodsol (p. 110)
Eng. & Wales	Humuspodzol
France	Podzol humo-ferrugineux (p. 51)
Neth. '50	Podzolised heath soil, mapping unit 48 (p. 22)
Neth. '61	Non-calcareous, high podzol, very poor not loamy fine sand; mapping unit 116
Neth. '66	'Haar' podzol soil (p. 132, 133, 134 and 190); mapping unit Hd21

Location. Gelderland Province, about 15 km northwest of Arnhem (see Fig. 1).
Parent material. Cover sand, i.e. an aeolian deposit from Weichsel (Würm) age.
Topography and elevation. On a fossil dune ridge about 5 m above a glacial outwash plain gently sloping to the southwest; about 40 m above sea level.
Drainage and ground water. No external drainage, some lateral water movement over the thin iron pan after showers. This kind of internal runoff gives rise to small intermittent ponds in relatively low places; just beyond the crest of the ridge there is such a wet spot, dry during sampling of this soil. The level of the ground water is most probably in excess of 10 m depth.
Present land use. Waste land with mainly ling (*Calluna vulgaris*), few volunteer Scots pine (*Pinus sylvestris*) and birches; on the above-mentioned wet spot, cross-leaved heath (*Erica tetralix*) and purple moor-grass (*Molinia caerulea*).
Range in land use. A small area with these soils is still unused carrying a *Genisto pilosae-Callunetum,* but most is afforested, chiefly with Scots pine (*Pinus sylvestris*) and smaller area of Douglas fir (*Pseudotsuga menziesii*) and larch (*Larix Kaempferi*). This is poor arable land with only a small area cultivated for rye and potatoes, although recently the practise of long-term leys has increased.

Discussion

This is the standard podzol of the sandy heaths of northwestern Europe, unreclaimed and with deep ground-water table. It must be emphasized that *these soils are never loamy, but are always sandy*. The podzolic and derno-podzolic soils in the USSR which are developed from loamy moraine material or loess–like material have no resemblance to this soil, but have more in common with the West European pseudogley soils (Kauritsjeva & Gromyko, 1974, e.g. Table 13 and 16; Gerasimov & Zonn, 1971; Targulian, 1974, p. 4-12).

The morphological characteristics of this soil meet the non-qualitative criteria for podzolic soils on the Soil Map of Europe and of the Podzol soils in *Soils and Men*.

The diagnostic criteria in Soil Taxonomy, the FAO and the French system justify its classification as a Humod, a Humic Podzol and a *podzol humo-ferrugineux* respectively.

These soils mostly have a thin iron pan directly below the B2h, which is often overlooked (careful cleaning of the profile with a knife may be helpful). It is never thicker than 1-2 mm and is not always continuous. Many undisturbed ignition samples such as the inset on the plate were prepared, the lower boundary of the white sand was always very abrupt, and the upper part often turned slightly reddish. Guy D. Smith (SSS, 1975, pedon 110, p. 703) describes a Belgian soil similar to this, which he classifies as a Placohumod. According to my opinion the thin iron pan in this kind of soil in the Netherlands is too weakly developed to satisfy the concept of the placic horizon (SSS, 1975, p. 33). If this soil has to be classified as a Placohumod, there will be hardly Haplohumods in the Netherlands. The placic horizon of the Placaquods of the United Kingdom (Ragg & Clayden, 1973, p. 117) and of the *Schwartzwald* in Germany (Mückenhausen et al., 1977, p. 112; Mückenhausen, 1975, p. 449) are a feature of upland soils and are more conspicuous.

It can be seen from the inset on the plate that this soil has practically no iron in the upper part of the B, yet it is an *Eisenhumuspodsol* in the German system. It is stated that 'the B1 horizon shows illuviation of predominantly brown-iron and organic material' (Mückenhausen et al., 1977, p. 110) and such a soil has a Bsh1 which is described as 'Ortstein mit Humus und Sesquioxiden' (Mückenhausen, 1975, p. 448 and Tafel 15).

Because of the upper Bh horizon I classified this soil as a humuspodzol, but according to Avery (personal communication, 1978) it is a humo-ferric podzol having also a Bhs horizon.

The B2h satisfies the stipulation of the Soil Survey Manual: 'Outstanding accumulation of decomposed organic matter . . . making it B2h' (SSS, 1951, p. 182), in the new definition of the subscript *h* (SSS, 1975, p. 462) the word 'outstanding' has been dropped. Due to the omission of this word in both new definitions, the B horizons have to be coded: B2h-B2hir-B3hir. The same distinctions are possible in some other systems: B2h-B2hs2-B2hs3 (Arbeitsgemeinschaft Bodenkunde, 1971, p. 29; FAO, 1974, p. 22 and 23); B2h-B2hfe-B3hfe (Jamagne, 1967, p. 36 and 37). Avery (1973, p. 332) uses Bh (with little iron) and Bhs (with more iron) 'diagnostic at group or subgroup level', but in the official British system of soil horizon nomenclature the Bh is a 'B-horizon containing translocated organic matter, associated with aluminium, or iron and aluminium, . . .' (Hodgson, 1974, p. 77).

This soil was sampled before the method of taking undisturbed ignition strips (Van der Voort, 1972) was developed. Previously the difference in iron content between the B2h and the B22 was known, but not that the iron content at the boundary between the A2 and the B2h was greater. The relatively high iron content of the B2h (0.13% Fe_2O_3) might be explained by this phenomenon. During the excursion on the occasion of the meeting in Stuttgart in 1971 (Pseudogley & Gley) Prof. Dr. G. Roeschmann demonstrated a *Calluna-Humuseisen-Podsol* on the Lüneburger Heide. Most probably the high iron content of the B2h of this soil, 1.13% Fe_d, (Hugenroth, 1971, p. 221), might be explained as contamination by the thin iron pan. This was not in the profile description, but observed by this author during that excursion.

Profile description

A1	0-5 cm	Black (10YR3/2) fine sand; strongly bleached, somewhat concealed by the organic matter; single grain to very weak, very fine subangular blocky; loose to very friable; abrupt, smooth boundary.
A2	5-18 cm	Grey (10YR5/1) fine sand, strongly bleached, the typical 'Bleicherde' (bleached earth) of the north-west European podzols; single grain; loose; abrupt, smooth, locally irregular when 'pockets' of A2 go down to 50 cm depth (not on the picture).
B2h	18-23 cm	Black (7.5YR2/1) fine sand, organic matter has filled in nearly all pores, after ignition this illuvial horizon turns white; massive; firm, hard; on the boundary between the black and the brown B a thin (1-2 mm) iron pan (2,9% Fe_2O_3), being the boundary between the iron-eluvial and the iron-illuvial layer; very abrupt, smooth boundary.
B22	23-50 cm	Dark reddish brown (5YR2/3) fine sand, coated with humus and iron (turns red on ignition), many wave lamellae with thicker humus coatings; in situ rather massive, when removed it crushes very easily to single grain; loose; diffuse, smooth boundary formed by the:
B3	50-90 cm	Very gradual transition, humus coatings becoming thinner with depth, thus exposing the yellowish colour of the iron coatings.
C	90-120 cm plus	Light yellowish brown (10YR6/4) sand; single grain; loose; the sand grains owe their yellowish colour to thin iron coatings (turns red on ignition).

The whole soil, both solum and substratum, is non-calcareous.

Analytical data

Hori-zon	Depth (cm)	Fe_2O_3 (%)	pH-KCl	Organic fraction				Particle-size distribution of mineral fraction (% of fine earth)				
				O.m. (%)	C(%)	N(%)	C/N	< 2	2-50	50-105	105-150	> 150 µm
A1	0-5	0.18	3.6	10.1	4.2	0.17	24.7	2	4.5	10.5	20	63
A2	10-15	0.02	3.8	1.7	n.d.	n.d.	n.d.	0.5	3.5	11.5	20	64.5
B2h	18-20	0.13	3.9	9.2	4.1	0.18	22.7	0	5.5	10	19	65.5
B22	25-30	0.65	4.4	3.3	n.d.	n.d.	n.d.	1.5	2.5	11.5	20	64.5
B3	50-75	0.32	4.6	0.6	n.d.	n.d.	n.d.	0	1	10	20	69
C	90-100	0.32	4.6	0.5	n.d.	n.d.	n.d.	0	1.5	10	19	69.5

Detailed particle-size distribution of the > 150 µm fraction of the C sample:

150-210	210-300	300-420	420-600	600-850	850-1200 µm
38	22	7	2	0.4	0.1

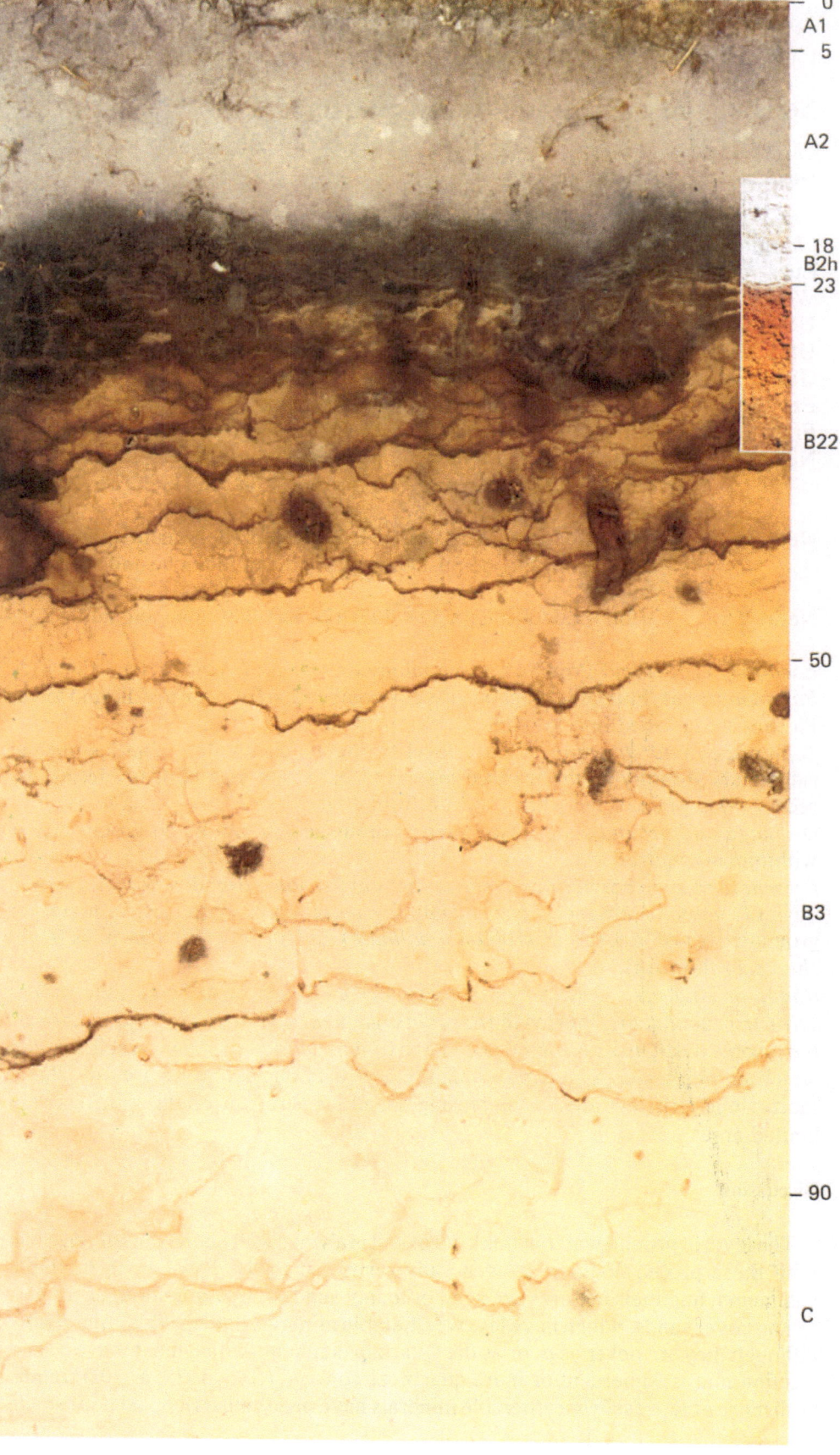

0	
A1	
5	
A2	
18	
B2h	
23	
B22	
50	
B3	
90	
C	

Soils of the Pleistocene sandy district: P4

General data

Classification

Eur.	Podzolized soil (p. 5 and 6)
World	Leptic Podzol (p. 39), coarse textured, gently undulating (p. 5)
USA '38	Brown Podzolic soil (p. 998, 1029 and 1030)
USA '75	Sandy, siliceous, mesic Entic Haplorthod (p. 346)
Ger.	Rostbraunerde (p. 94)
Eng. & Wales	Typical (non-humic) brown podzolic soil
France	Sol ocre podzolique modal (p. 52)
Neth. '50	Sandy push moraines complex; mapping unit 51, called brown forest soil or brown podzolic forest soil on sand in the book (p. 19 and 20)
Neth. '61	Non-calcareous sand soil, high, podzol, predominantly poor sand, complex of the ice-pushed ridges, gravelly loamy sand; mapping unit 123
Neth. '66	'Holt' podzol soil (p. 122, 123, 124 and 188); mapping unit gY30

Location. Gelderland Province, some kilometres west of Arnhem (see Fig. 1).
Parent material. Fluviatile, slightly gravelly coarse sand from the middle and lower Pleistocene, pushed into low hills by the Saale ice sheet.
Topography and elevation. Nearly level to slightly undulating. The ice-pushed ridge ends abruptly a few hundred meters to the south as a bluff bordering the recent flood plain of the Rhine, a drop from 40 m to 10 m above sea level.
Drainage and ground water. No external drainage, no runoff; no ground water within many metres.
Present land use. Former oak coppice, not cut for some decades, forestry nowadays of minor importance, used mainly for recreational purposes. A luxuriant ground cover of bracken (*Pteridium aquilinum*) and some wavy hair-grass (*Deschampsia flexuosa*).
Range in land use. Used almost exclusively as forest, with Scots pine (*Pinus sylvestris*) dominant, Douglas fir (*Pseudotsuga menziesii*) and larch (*Larix kaempferi*), some oak (*Quercus robur*) and beech (*Fagus sylvatica*). Some arable land on which potatoes and rye form the main crops with an increasing area of silage-maize; long-term leys are also increasing. Heathlands remain on this kind of soil in limited areas.

Discussion

The level of ground water in this kind of soil is always well below the solum. Soils with these profile features are generally restricted to the Pleistocene fluviatile sediments, and their textures are sand or loamy sand. They are seldom encountered on cover sands, which have less weatherable minerals than the fluviatile sands, although they are never as poor as the Quartzipsamments which must 'have a sand fraction that is 95 percent or more quartz, zircon. . .' (SSS, 1975, p. 202). Sands with more or less easily weatherable minerals have been called rich and poor sands in the local field jargon (poor and very poor sands respectively in the legend of the Soil Map of the Netherlands, scale 1:200 000). This distinction is drawn when more than 15% easily weatherable minerals occur in the 50-100 µm separate. Contrary to earlier opinions it is now thought that the slight difference in mineralogy between for example soil P3 (p. 165) and this soil, is more important as a soil-forming factor than the vegetation (Edelman, 1963).

The absence or near-absence of a typical podzol (=ash-like) eluvial horizon in this soil is the cause of the continuing discussion about whether or not this soil is a podzol. In *Soils and Men* it has been stated: 'Essentially, the Brown Podzolic soil is an imperfectly developed Podzol. . .' (USDA, 1938, p. 1029), this is reflected by the adjective 'Entic' in the modern name: Entic Haplorthod, indicating they are intergrading to the Entisols. Many of the Brown Podzolic soils in the '38-system today are Dystrochrepts, and some are Udipsamments, e.g. the Plymouth soil series: the soil from slide 3-5 of the Marbut Memorial Slides (SSSA, 1968, p. 5) resembles this Dutch soil under discussion.

Another soil that looks identical to this soil, is soil 43 in the Soil Atlas of the USSR (Kauritsjeva & Gromyko, 1974); such soils are called 'weakly sod-podzolized sandy soils'.

The Germans considered this soil to be an intergrade between a *Braunerde* and a *Podsol*: a *Podsol-Braunerde* (Mückenhausen et al., 1962, p. 80), but in the second edition of their system they call it a Rostbraunerde (Mückenhausen et al., 1977, p. 94 and 112). The English use the adjective podzolic and the French *podzolique*. The FAO system names this soil a Leptic Podzol, from Gr. *leptos,* shallow (FAO, 1974, p. 19). The conclusion is that no one classifies this soil whole-heartedly as a podzol.

The so-called concealed A2 is coded A1(2) which will be the same in both the old and new USA systems (SSS, 1951; SSS, 1975) and also in the French system (Jamagne, 1967). In the FAO system it will be an AhE (FAO, 1974) and in the new British Soil Survey Field Handbook such 'a dark surface horizon with uncoated sand and silt particles' (Hodgson, 1974, p. 74) is indicated as an Ah/Eh, whereas in the German system a 'mineral soil horizon, which in general is coloured nearly homogeneous by acid organic substances and contains mostly bleached quartz grains' is coded as an Aeh (Arbeitsgemeinschaft Bodenkunde, 1971, p. 29).

The B2 horizon is called B2, a B2irh, a B2h/fe, a Bsh1 (twice) and a Bv1 (same systems in the same sequence as referred to above). All these codes point to illuviation, except the German Bv which indicates 'a horizon between the A and the C horizon which is browned and siltified by weathering without or only with negligible illuviation' (Arbeitsgemeinschaft Bodenkunde, 1971, p. 29, item 6 on p. 113). This clearly illustrates the different opinions about the genesis of unsaturated, sandy brown soils with bleached sand grains in the upper horizon!

The difference between this fluviatile sand and the aeolian sand of the previous soil is striking. The latter is well-sorted, i.e. a high 'peak' in the 150-210 µm separate and a low and short 'tail' to the coarse side; the fluviatile sand has a 'flat' distribution curve and a long tail to the coarse side.

Profile description

A0	+2-0 cm	Un-decomposed and partly decomposed litter, mainly derived from oak leaves; abrupt, smooth boundary.
A1(2)	0-6 cm	Very dark grey (10YR2.5/1) loamy sand to sandy loam; single grain, loose to very friable; many sand grains bleached, but the A2-character is concealed because of the high organic matter content.
AB	6-20 cm	Transitional horizon, characterized by a decreasing number of bleached sand grains, and a gradual change of colour from greyish to brownish shades.
B2	20-45 cm	Dark brown (10YR3.5/3) loamy sand; single grain to very weak fine angular blocky; loose to very friable; sand grains with iron coatings; the fine material between the sand grains consists of mineral silt particles and organic mull-like moder.
B3	45-80 cm	Very gradual transition to the subsoil, becoming lighter in colour and lower in organic matter with depth.
C	80-120 cm plus	Yellowish brown (10YR5/4) sand; single grain; loose; colour is due to a thin coating of iron compounds.

The soil is non-calcareous throughout.

Analytical data

Horizon	Depth (cm)	Fe_2O_3 (%)	pH-KCl	Organic fraction				Particle-size distribution of mineral fraction (% of fine earth)				
				O.m. (%)	C(%)	N(%)	C/N	< 2	2-50	50-105	105-150	> 150 µm
A1(2)	0-6	0.41	3.0	10.0	5.3	0.33	16.1	5	15	6	10	64
B2	25-35	0.69	4.0	3.4	1.6	0.07	22.9	5	16	7	10	62
B3	50-60	n.d.	4.3	1.7	n.d.	n.d.	n.d.	4.5	15	8	11	61.5
C	90-100	0.52	4.7	0.3	n.d.	n.d.	n.d.	1.5	10	4	8	76.5

Detailed particle-size distribution of the > 150 µm fraction of the C sample:

150-210	210-300	300-420	420-600	600-850	850-1200	1200-1700 µm
15	18	21	12	7	3	0.5

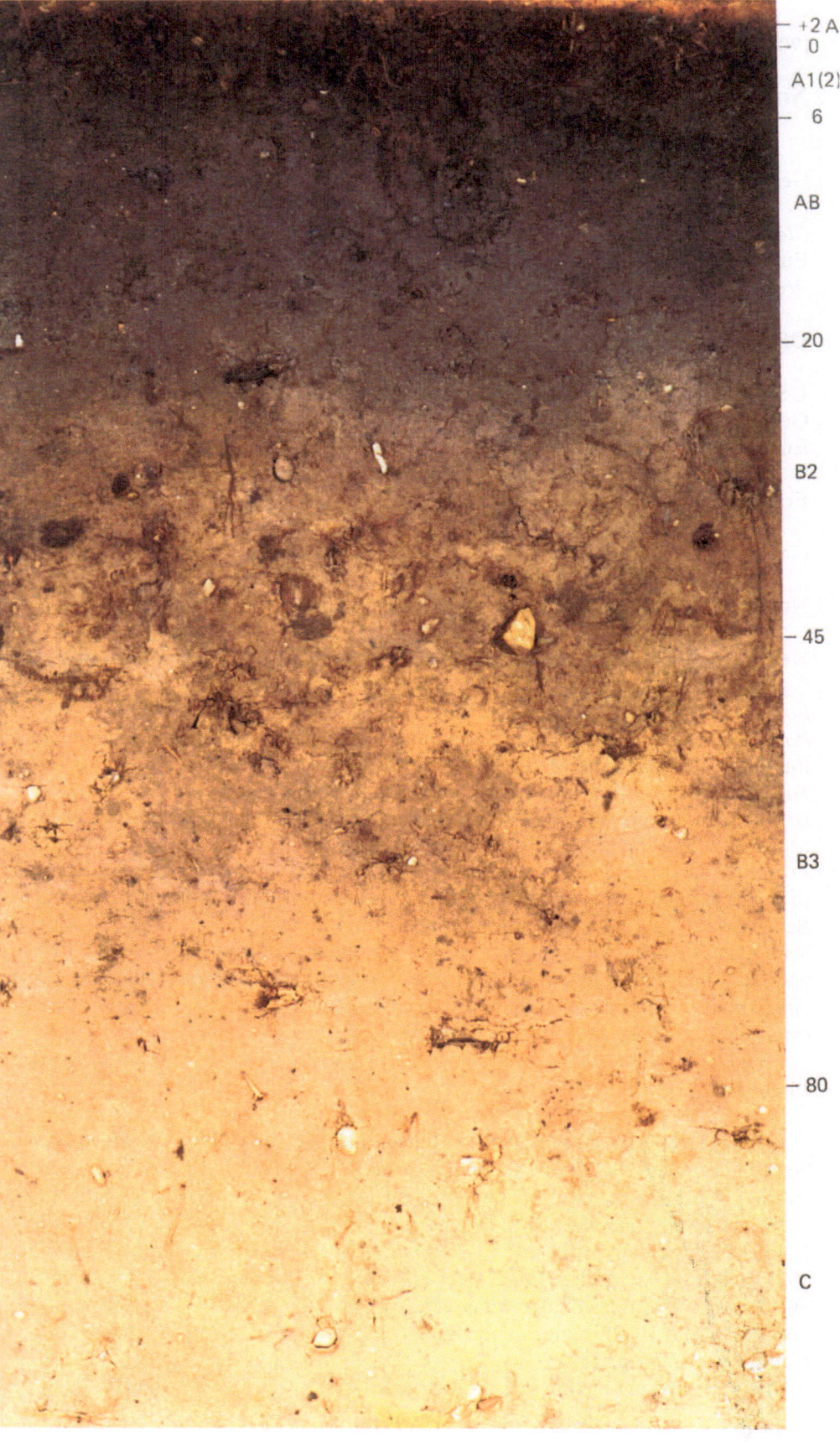

+2 A
0
A1(2)
6

AB

20

B2

45

B3

80

C

Soils of the Pleistocene sandy district: P5

General data

Classification

Eur.	Podzolized soil (p. 5, 6 and 44-47)
World	Not mentioned specifically, is a special kind of Leptic Podzol, coarse textured, level (p. 5)
USA '38	Not mentioned specifically, is a special kind of Brown Podzolic soil (p. 998, 1029 and 1030)
USA '75	Sandy, siliceous, mesic Ultic Haplorthod (p. 346)
Ger.	Not mentioned
Eng. & Wales	Not mentioned specifically, is a special kind of typical (non-humic) brown podzolic soil
France	Not mentioned specifically, is a special kind of a sol ocre podzolique
Neth. '50	Not mentioned, included in mapping unit 22, brown river 'loam' soils (p. 60-64)
Neth. '61	Not mentioned specifically, included in mapping unit 79, high older river clay soils
Neth. '66	'Horst' podzol soil (p. 121, 122, 187 and 188); mapping unit Y23b

Location. Limburg Province (see Fig. 1).
Parent material. Cover sand, i.e. aeolian sand of the Weichsel age, overlying fluviatile sand from the Meuse of about the same age.
Topography and elevation. Level; about 25 m above sea level.
Drainage and ground water. No external drainage, no runoff, ground-water level at 4-5 m below the surface.
Present land use. The soil was sampled from the wall of a sand pit in a waste corner of land between two road-forks, with a vegetation of ling (*Calluna vulgaris*) and Scots pine (*Pinus sylvestris*) in the midst of arable land.
Range in land use. Predominantly arable, the main crop is spring barley, with rye and oats, and winter wheat on the loamy variants, also potatoes, sugar beet and silage-maize. Some horticulture: asparagus; some orchards: apples and cherries (morellos). A small area of forest, mainly Scots pine.

Discussion

Although plate P5 is not a picture of a soil but of a subsoil, it is included in this book to illustrate the phenomenon of an 'argillic horizon composed of lamellae' (SSS, 1975, p. 25). It has also been named textural subsoil lamellae (Dijkerman, 1965), or simply lamellae (MacVicar et al., 1977, p. 109), banded B horizon (De Bakker & Schelling, 1966, p. 177), gebänderter B-horizont (Mückenhausen et al., 1977, p. 86 and 96; Müller, 1967, Tafel 114), Bänder (Ehwald et al., 1966, p. 41 and 50), Tonanreicherungsbänder (Paas, 1961), kovárvány (Stefanovits, 1971, p. 71-77, 106-107, photo III-2), bandes (Jamagne, 1967, p. 32), smugi żółtobrunatne (Kuźnicki et al., 1974, p. 36).

The upper 60 cm of this soil (not shown) is analogous to the preceding soil, a

Leptic podzol, Brown Podzolic, Haplorthod, and so on.

Lamellae are formed exclusively in sandy soils, and in sandy subsoils below loess, glacial till or pleistocene loamy river sediments. Both lamellae and inter-lamellae sandy layers are non-calcareous; if there is a calcareous subsoil, the lower boundary of the lowest lamella coincides with the depth of decalcification (Müller, 1967, Tafel 114; Paas, 1961, p. 171). Lamellae are never found below the ground-water level, nor in the fluctuation zone, but if the ground water is very deep, they may also be found at a depth of some metres. Some suppose the lamellae formation dates from a time of periglacial environment (Kuźnicki et al., 1974, p. 36), but there is evidence of more recent development. Goetz (1970, p. 64 and 68) studied lamellae that were 760 years and also 12 000 years old.

Most systems of soil classification have no special category for a soil with lamellae in the subsoil. The German system only mentions them in the discussion about *Parabraunerden aus Sand* (Mückenhausen et al., 1977, p. 86 and 96), but in the classification system they do not form a separate type or subtype. Hugenroth (1971, p. 215) calls such a soil a *Bänderparabraunerde, schwach podsoliert.*

Only the new American system has several subgroups with lamellae in the subsoil, some subgroups occur in the Psamments and some in the Haplorthods, which are called either Alfic or Ultic. The distinction relies on base saturation (more or less than 35% at a depth of 1.25 m below the uppermost lamella) and on the temperature regime (frigid versus mesic or warmer). The definition is such that soils with a frigid temperature regime are Alfic; mesic soils are Ultic only when the base saturation is low (SSS, 1975, p. 206 and 346). Sandy podzolized soils always have a low base saturation throughout (at least in northwestern Europe) and in the Netherlands the soil temperature regime is decisive for their classification. All Dutch soils are mesic and all Dutch soils with lamellae are acid, so this soil is an Ultic Haplorthod.

The lamellae are clearly thin Bt horizons, formed by clay illuviation. The inter-lamellar sandy layers are called C horizons; they might also be coded B3 horizons between B2 horizons (SSS, 1975, pedon 19, p. 523) or A2 horizons (Wurman et al., 1959), thus indicating the divergence of opinion of pedologists: illuviation from the upper part of the solum or mainly from the inter-lamellar layers.

The boundary between the aeolian cover sand and the fluviatile sand is at 145 cm depth. There is no 'significant change in particle-size distribution or mineralogy' (SSS, 1975, p. 459) and the small difference does not justify the use of the prefix II in the codes of the lowest horizons.

The lamellae have a greater clay content and also a greater iron content than the inter-lamellar layers.

The detailed particle-size analyses show the difference between the aeolian sand (the C13) and the somewhat coarser fluviatile sand (the Bt3).

Profile description

	0-60 cm	Not on the plate, soil horizons and colours as soil P4.
B3	60-65/70 cm	Dark yellowish brown (10YR4/4) fine sand; single grain; loose; sand grains with thin iron-coatings, the fine material between the grains consists of silt and mull-like moder; gradual, smooth to wavy boundary.
C with Bt-lamellae	65/70-180 cm plus	The interlamellae material is light yellowish brown (10YR6/4) fine sand; single grain; loose; sand grains with thin iron-coatings. The lamellae consist of strong brown (7.5YR5/7) loamy fine sand to fine sand; weak, angular blocky to massive; slightly hard; oriented clay-coatings on the sand grains, partly as bridges between the grains; upper transition very abrupt, lower transition clear; both smooth.
		In the middle part of the plate distinct, common, medium mottles, being strio-tubules (caused by burrowing grubs of cockchafers or dung-beetles).

The whole soil is non-calcareous.

Analytical data

Horizon	Depth (cm)	Fe_2O_3 (%)	pH-KC1	O.m. (%)	Particle-size distribution of mineral fraction (% of fine earth)				
					< 2	2-50	50-105	105-150	> 150 μm
B3	60-70	0.74	4.3	0.5	1.6	15.8	19.5	34.5	28.6
C11	70-80	0.57·	4.4	0.3	0.3	8.8	19.3	31.8	39.8
Bt1	85-90	1.85	3.8	0.3	6.6	8.8	12.9	35.4	36.3
C13	125-135	0.71	4.1	0.2	0.8	5.3	19.8	48.0	26.1
Bt3	145-150	1.38	4.0	0.2	5.3	4.1	6.3	32.8	51.5

Detailed particle-size distribution of the > 150 μm fraction of the C13 and Bt3 samples:

150-210	210-300	300-420	420-600	> 600 μm
23.6	2.2	0.3	0	0
36.1	11.0	3.0	0.9	0.5

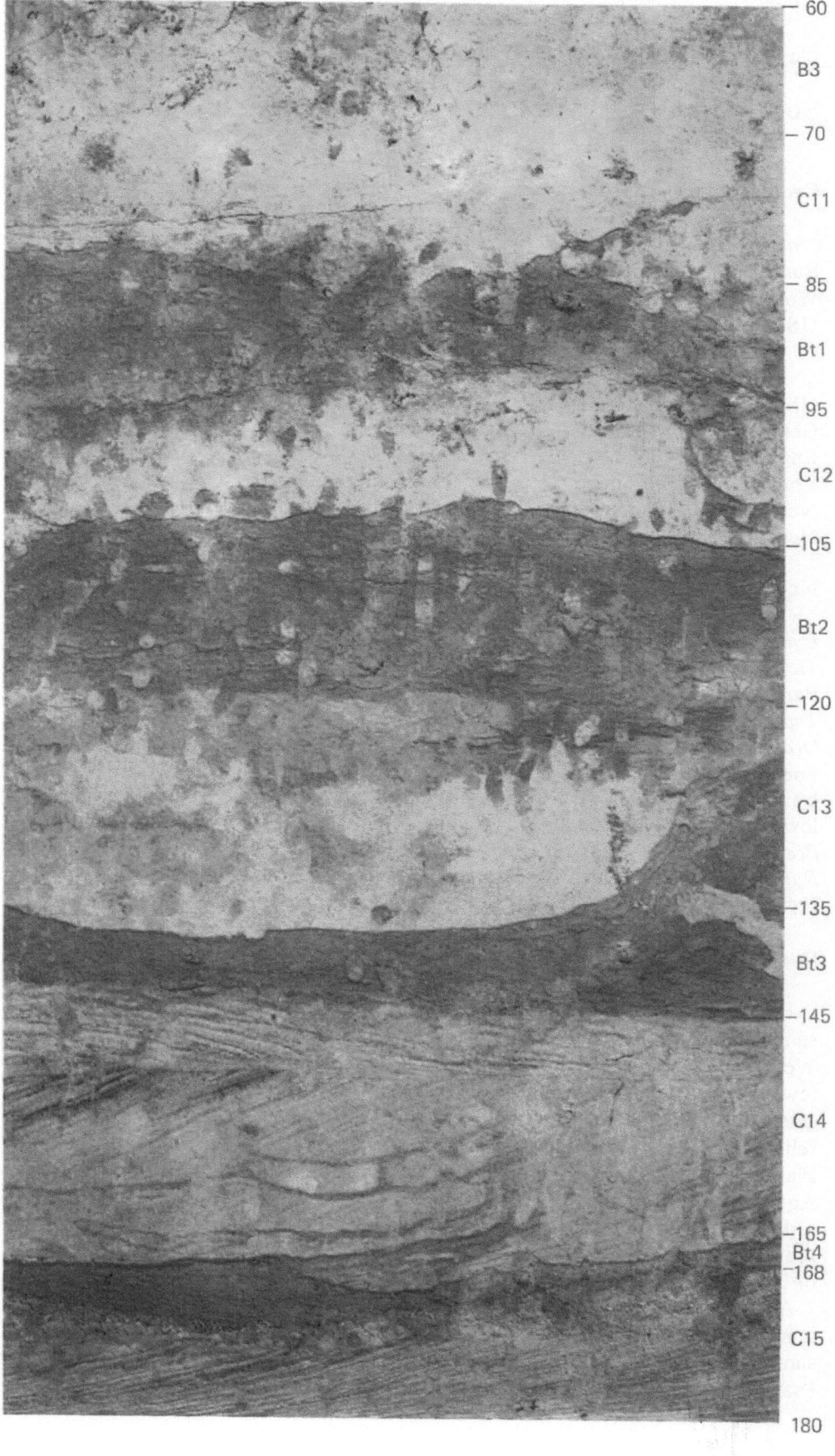

Soils of the Pleistocene sandy district: P6

General data

Classification

Eur.	Not mentioned as such in the list of soil units
World	Humic Gleysol (p. 33), coarse textured, level (p. 5)
USA '38	Called Wiesenboden (Meadow) in 1938 (p. 999), described toge-ther with Ground-Water Podzols and Half-Bog soils on p. 1110, in 1949 called Humic-Glei soil (p. 119)
USA '75	Sandy, siliceous, mesic Typic Humaquept (p. 243 and 244)
Ger.	Typischer Gley (p. 151)
Eng. & Wales	Typical humic-sandy gley soil
France	Sol humique à gley, à anmoor calcique (p. 75)
Neth. '50	Moist sandy soil, mapping unit 39; called black forest soil (p. 27)
Neth. '61	Non-calcareous low sand soil, gley soil, very poor sand, loamy fine sand; mapping unit 102
Neth. '66	Black 'beek' earth soil (p. 149-151 and 193); mapping unit pZg23

Location. Utrecht Province, about 20 km southeast of the town of Utrecht (see Fig. 1).

Parent material. Cover sand, i.e. aeolian sand from the Weichsel age.

Topography and elevation. Level; about 5 m above sea level.

Drainage and ground water. No tile drains, ditches spaced 50-75 m apart, shallow open field drains spaced 15 m apart, all draining via an irregular network of ditches by gravity into a small river which discharges into Lake Yssel. The ground-water level fluctuates between 40 and 120 cm.

Present land use. Grassland.

Range in land use. Nearly all these soils are used as grassland, the time of reclamation is unknown, possibly as long ago as the plaggensoils (soils A1 and A2).

Discussion

Both this soil and the hydromorphic podzol soil (p. 157) are developed from the same kind of parent material: Pleistocene aeolian sand, deposited at the end of the Weichsel glacial. Presumably these sands were initially calcareous but contained few weatherable minerals. This sand blankets older Pleistocene deposits, such as glacial till and fluviatile coarse sediments; for this reason it is called cover sand. Its relief is characterized by low ridges, carrying soils like the Humod of p. 165, or the Plaggepts of p. 137 and 141; rather level or depressional areas, with soils like the Aquod of p. 157, and very shallow valleys with soils similar to this soil. During sedimentation of the cover sands these valleys drained the snow melt-water into the rivers, nowadays the brooks in these valleys form the actual drainage courses of the area (Fig. 31).

After amelioration of the climate the permafrost disappeared, the vegetation increased in density and the sand transport ended. At that time all soils on cover sands had little horizon differentiation and were either Typic or Aquic Cryo-Psamments.

The decisive factor of soil formation which causes the differences between this soil and the hydromorphic podzol soil (p. 157) is *relief*. Hydromorphic podzols are situated on level or in closed depressional areas, whereas the Humaquepts are almost exclusively restricted to the valleys. In the hollows all excess of precipitation has to infiltrate into the soil and water is constantly moving downwards. In the valleys there is a constant supply of water from elsewhere, both laterally and by seepage from upward moving ground water (Schelling & Marsman, 1973, Fig. 6). As a consequence this soil is continuously enriched, and the podzol site impoverished.

The valleys were reclaimed some centuries ago from alder forest and utilized as grassland. The podzol area originally carried a birch forest with some oak and purple moor-grass on the wet places and oak was dominant on the sites with deeper ground-water levels. As the use of these areas was intensified, they gradually changed into heathland (cross-leaved heath, purple moor-grass and bog-myrtle on the wet places, and predominantly ling on the higher places). Most of the heathland was reclaimed as grassland, arable or forest in recent times (Fig. 30).

The classification of this soil produces some difficulties in the French and in the recent USA system. Liming and fertilizing the dark topsoil made it less acid than it was initially. Most probably in the French system the soil was originally a *sol humique à gley à anmoor acide*; now it is changed into an *anmoor calcique*. The distinction between the mollic and the umbric epipedon in Soil Taxonomy is a base saturation of more than 50%, these kind of soils have values between 40 and 60%. Both dark surface horizons have to be thicker than 25 cm in sandy soils (SSS, 1975, p. 16, item 5d1). Therefore this soil could also be classified as a Typic Haplaquoll, a Mollic Psammaquent or a Humaqueptic Psammaquent, depending on the base saturation (over or under 50%) and the thickness (over or under 25 cm) of the epipedon. The soil is classified as if having a base saturation under 50% and a thickness of the epipedon over 25 cm.

Clearly this soil was formerly ploughed, but in the upper part of the former plough layer (Apg) a new turf has been developed (A11g), whereas the marked transition to the C1g is being blurred by burrowing earthworms. The two lowest horizons would have to be coded Go and Gro in the German system, Cg and Cgr in the FAO system, Cg and Cg/CG in the British system. In the French system both horizons have to be coded CG (g is restricted to pseudogley phenomena), and in the new USA system both horizons have the same code Cg (cf. p. 83).

The difference in organic-matter content between the newly developed turf and the former plough layer is conspicuous: 12.2% against 3.7%. The C-N ratios of this soil are much lower than those of the hydromorphic podzol soil P1 (p. 156).

Profile description

A11g   0-5 cm   Black (10YR2.5/1) loamy fine sand; turf of grassland developed in an old
                         plough layer; common, distinct, medium strong brown (7.5YR5/6) mottles;
                         friable; abrupt, smooth boundary.
Apg   5-25 cm   Very dark grey (10YR3/1) loamy fine sand; common to many, distinct,
                         medium reddish yellow (7.5YR6/6) mottles; friable; in the lower part many
                         fine 'krotovinas' filled in with C-material; abrupt, somewhat irregular
                         boundary (plough or spade traces).
C1g   25-90 cm   Light grey (2.5Y6/1) loamy fine sand; common, prominent to distinct,
                         medium strong brown (7.5YR5/6) mottles; in the upper part many fine
                         'krotovinas' filled in with A1-material; in the lower part clearly stratified;
                         gradual, irregular boundary.
C2G   90-120 cm plus Grey (10Y5/1) fine sand; few, faint to distinct, medium yellowish brown
                         (10YR5/5) mottles; transitional horizon to the G-horizon with neutral co-
                         lours and no brown mottles.

Analytical data

Hori-zon	Depth (cm)	CaCO₃ (%)	pH-KC1	Organic fraction				Particle-size distribution of mineral fraction (% of fine earth)					
				O.m. (%)	C(%)	N(%)	C/N	< 2	2-16	16-50	50-105	105-150	>150 µm
A11g	0-5	0.2	6.3	12.2	4.6	0.52	8.8	5	2.5	16	24	17	36
Apg	10-23	0.1	5.5	3.7	2.1	0.18	11.7	5	3.5	15	25	14	38
C1g	38-48	0.1	5.5	0.4	n.d.	n.d.	n.d.	1.5	1.5	16	38	22	21
C2G	105-115	2.7	8.0	0.3	n.d.	n.d.	n.d.	4.5	0.2	3.7	23	32	37.7

Detailed particle-size distribution of the > 150 µm fraction of the C2G sample:

150-210	210-300	300-420	> 420 µm
25	9	2.5	1.2

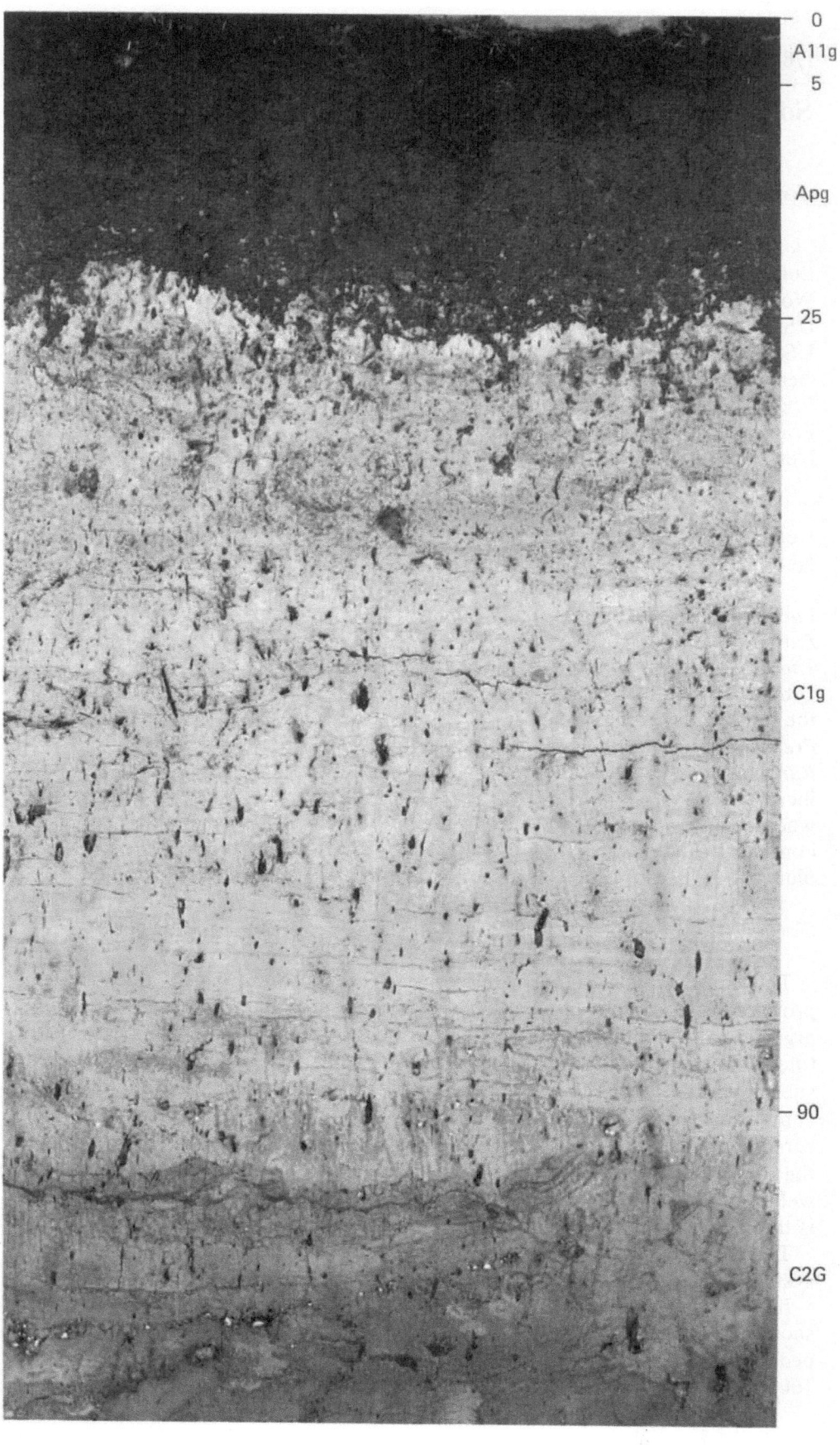

Soils of the loess district: L1

General data

Classification

Eur.	Gray-brown podzolic soil (p. 4, 5 and 54-56)
World	Orthic Luvisol (p. 38), medium textured, gently undulating (p. 5)
USA '38	Gray-Brown Podzolic soil (p. 973, 998, 1033 and 1034)
USA '75	Fine-silty, mixed, mesic Typic Hapludalf (p. 129 and 130)
Ger.	Basenreichere Parabraunerde (p. 98)
Eng. & Wales	Typical argillic brown earth
France	Sol lessivé modal (p. 47)
Neth. '50	Not recognized as such, called simply loess loam; mapping unit 52, referred to in the book as 'slightly podzolised brown forest profile' (p. 172)
Neth. '61	High loess soil with a textural B, loam; mapping unit 144
Neth. '66	'Rade' brick soil (p. 142 and 191); mapping unit BLd6

Location. Limburg Province, about 20 km north of Maastricht (see Fig. 1).
Parent material. Loess.
Topography. Gently undulating, about 55 m above sea level.
Drainage and ground water. No external drainage, some runoff during showers; the level of the ground water is always several metres below the surface.
Present land use. Arable land, wheat stubble.
Range in land use. Mainly arable, in the last few years the area of grassland has increased. Main arable crops: sugar beet, potatoes and wheat (mainly winter wheat), some spring barley, an occasional rye crop but silage-maize has increased considerably in importance. Orchards with apples, pears and to a lesser extent plums and morellos, some strawberries are the horticultural crops grown.

Discussion

This soil is typical of the scattered loess areas in northwestern Europe. The profile consists of a plough layer which is not very dark and which has a rather low organic-matter content, a yellowish subsurface horizon with a weak platy structure (these horizons often are absent as a result of sheet erosion) and a finer textured subsoil with weak but coarse prismatic structure (N.B. contrary to the common appearance in the field, with a smoothed profile face, the monolith has been carefully prepared in order to show these big prisms). Like the soil on the levee of the Rhine (soil F3, p. 101) this soil has many vertical earthworm burrows. It is well-drained, acid in the upper part, neutral in the lower part and if the loess deposit is thick enough, calcareous in the deeper subsoil.

There is little need for discussion about the classification of such a soil in most systems, because it is a typical example.

In the terms of Soil Taxonomy, there could be a discussion whether this soil should be an Agrudalf or a Hapludalf. Like the Belgian loess soil described as pedon 22 (SSS, 1975, p. 528 and 529) this soil has also 'been farmed for more than 1000 years' (SSS, 1975, p. 126), but its B horizon does not satisfy all requirements of the agric horizon (SSS, 1975, p. 27 and 28). It has many earthworm channels 'coated with a dark-colored mixture of organic matter, silt and clay', but there are no lamellae. The C-N ratio is higher than 8 and the pH is not 'close to neutrality'. This Dutch soil is not an Agrudalf but a Typic Hapludalf.

In the French system the non-hydromorphic soils with a textural B, the group of *sols lessivés* have four subgroups: brown, modal, acid and weakly podzolized. The *sol brun lessivé* must have an *indice d'entrainement* (illuviation index) between 1:2.0 and 1:1.4 (in this soil it is 1:2.1), in the other subgroups this index must be below 1:2.0. The main distinction between the modal and the acid subgroup is the pH of the B horizon, the limiting value being 5.5. It is not stated which analytical method has to be used, if it is pH-H_2O this soil is a *sol lessivé modal* (as the pH-KC1 might be up to one unit lower than the pH-H_2O), if not, it belongs to the acid subgroup.

On the highest level of the system of England and Wales this soil belongs to the brown soils: 'Soils, excluding pelosols, with weathered, argillic or paleoargillic B and no diagnostic gleyed horizon at 40 cm or less' (Avery, 1973). This major group is subdivided into 8 groups, one of them being the argillic brown earth.

All systems of horizon nomenclature discussed agree about the designation of the Ap and Bt horizons. The eluvial horizon is coded A2 in both the old and recent USA system, and also in the French and Dutch systems. It is E in the FAO system, Eb in the system of England and Wales and A3 in Germany (Mückenhausen et al., 1962, p. 43) recently being changed into A1 (l as in the French *lessivé*) (Arbeitsgemeinschaft Bodenkunde, 1971, p. 29).

The texture is normal for West European loess: a high silt content and a low sand content; the texture profile is typical for a soil with clay illuviation: first a gradual increase from 9% to 21% clay and lower down a decrease to 13%.

Loamy soils used as arable land always have low organic-matter contents in the Netherlands, whether they are derived from loess or from marine sediments (cf. the soils M4, p. 76 and M6, p. 84).

The decalcification depth (2.7 m) is normal for deep loess deposits which were originally calcareous and not eroded.

Profile description

Ap	0-20 cm	Dark grey brown (10YR4/2) silt loam; typical structure for arable land on loess in autumn just before ploughing: in the upper part firm dense clods and in the lower part nearly massive; abrupt and smooth boundary.
A2	20-35 cm	Dark yellowish brown (10YR4/4) silt loam; weak, thin to medium platy structure; friable; smooth to wavy boundary.
B1t	35-55 cm	Dark brown (7.5YR3/4) silt loam; weak, very coarse prismatic structure breaking to moderate, medium subangular blocky structure; in many pores and on some ped surfaces, in particular on the surfaces of the prisms, coatings of clay; firm; gradual, smooth boundary.
B2t	55-110 cm	Dark brown to brown (7.5YR4/4) silt loam; structure like in B1t but somewhat stronger graded (smoother ped surfaces, sharper vertices); conspicuous reddish brown (5YR4/5) clay coatings; firm; gradual, smooth boundary.
B3t	110-120 cm plus	Yellowish brown (10YR5/4) silt loam; very low grade of structure; still many clay coatings in the pores; friable; diffuse, smooth boundary to C1-horizon at 1.70 m.

Especially in the B-horizon some earthworm burrows plastered with dark wormcasts. At 2.70 m depth there is the lower boundary of the C1-horizon, very abrupt and smooth to the calcareous loess, in the upper part there is a fine lime segregation (pseudomycelium) over a few cm thickness.

Analytical data

Horizon	Depth (cm)	Organic fraction				Particle-size distribution of mineral fraction (% of fine earth)			
		O.m. (%)	C(%)	N(%)	C/N	< 2	2-16	16-50	> 50 μm
Ap	0-20	2.2	2.0	0.12	17	9	14	60	18
A2	20-35	0.8	n.d.	n.d.	n.d.	10	14	62	14
B1t	35-55	0.6	n.d.	n.d.	n.d.	18	14	55	13.5
B2t	55-110	0.6	n.d.	n.d.	n.d.	21	12	57	10
B3t	110-170	0.4	n.d.	n.d.	n.d.	17	9	64	10.5
C1	170-270	0.2	n.d.	n.d.	n.d.	14	12	65	9
C2	> 270	0.2	n.d.	n.d.	n.d.	13	11	68	8

Horizon	Fe_2O_3 (%)	$CaCO_3$ (%)	pH-KC1	Extractable cations (% of sum)					C.E.C. (meq./100 g)
				Na	K	Mg	Ca	H	
Ap	1.63	n.d.	3.8	0.0	0.0	1.2	9.6	89.2	8.3
A2	1.58	n.d.	4.0	0.0	0.0	1.5	21.5	76.9	6.5
B1t	2.63	n.d.	4.8	1.1	1.1	6.3	64.1	27.3	9.5
B2t	3.32	n.d.	5.0	1.7	2.5	9.3	66.1	20.3	11.8
B3t	2.78	n.d.	5.0	2.1	2.1	9.6	66.0	20.2	9.4
C1	2.51	0.1	5.8	2.1	1.0	9.5	80.0	7.4	9.5
C2	2.12	15.2	7.9	2.2	1.1	8.6	83.8	4.3	9.3

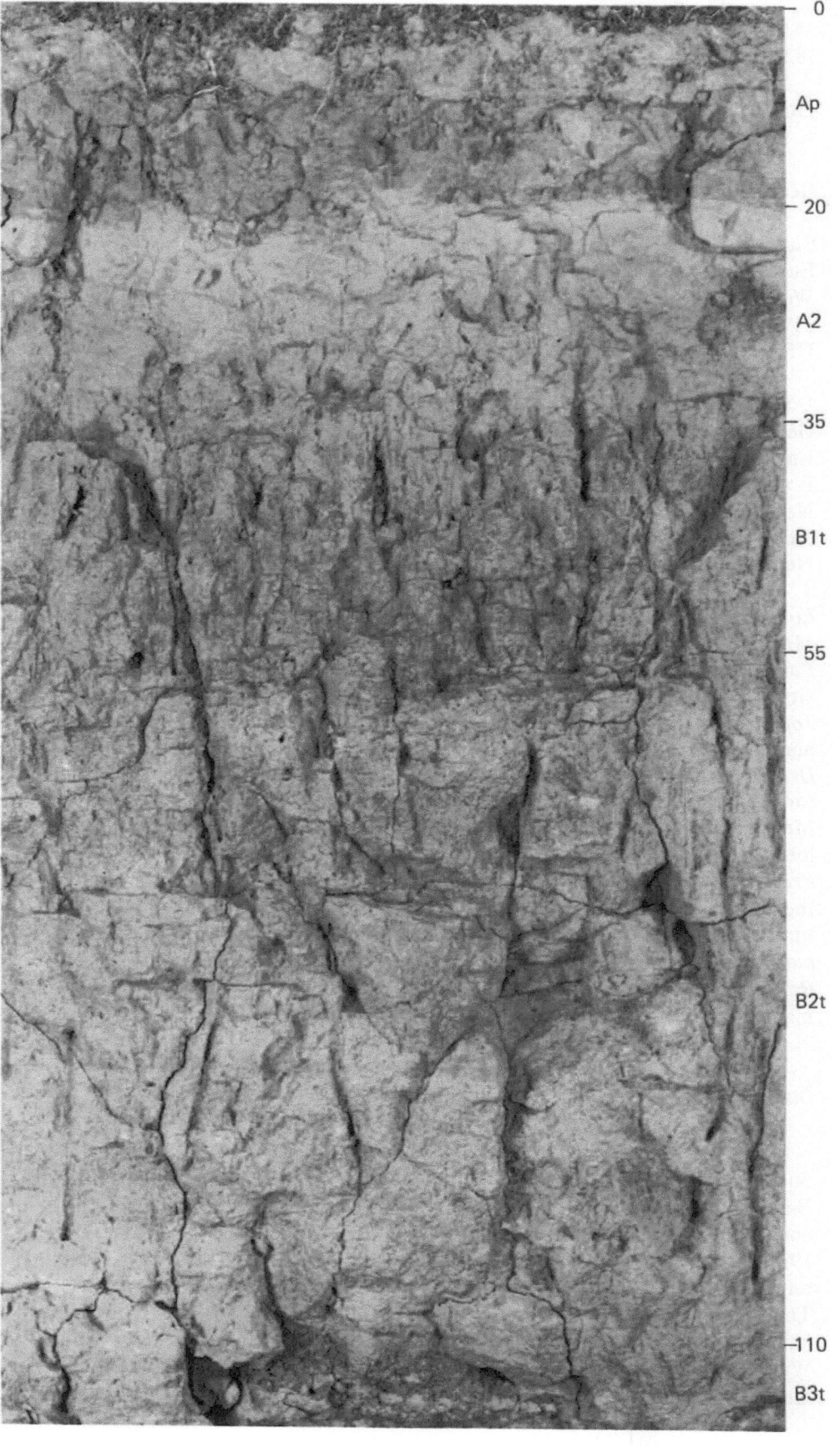

	0
	Ap
	20
	A2
	35
	B1t
	55
	B2t
	110
	B3t

182

Soils of the loess district: L2

General data

Classification

Eur.	Pseudogley soil (p. 5)
World	Gleyic Podzoluvisol (p. 39), medium textured, level to gently undulating (p. 5)
USA '38	Low-Humic Glei soil or Planosol according to the 1949 modification
USA '75	Fine-silty, mixed, mesic Typic Glossaqualf (p. 112-114)
Ger.	Typischer Pseudogley (p. 131) or Gley-Fahlerde (p. 102)
Eng. & Wales	Typical (argillic) stagnogley or typical argillic gley soil
France	Sol à pseudogley de surface (p. 76) or sol à gley lessivé (p. 75)
Neth. '50	Not mentioned
Neth. '61	Not mentioned
Neth. '66	'Kuil' brick soil (p. 136, 137 and 190); mapping unit BLn5g

Location. Limburg Province, in the Limbricht Forest, about 22 km northeast of Maastricht (see Fig. 1).
Parent material. Weichsel loess overlying sandy gravelly Pleistocene Meuse deposits.
Topography and elevation. Level; about 36 m above sea level and about 8 m above and at about 5 km distance from the first bottom of the Meuse.
Drainage and ground water. Shallow ditches alongside some of the dirt roads in the forest, draining into a small brook outside the forest which flows by gravity into the Meuse. The ground-water level fluctuates greatly, in summer it is well below the loess in the coarse river sediment, in winter it may be at the surface occasionally.
Present land use. Most probably it has never been cultivated (very exceptional for the Netherlands!). The actual vegetation, neglected forest, is now almost seminatural: oak, beech, pine and a luxuriant undergrowth of bracken (*Pteridium aquilinum*) and brambles (*Rubus fruticosus*).
Range in land use. Mainly deciduous forest, some grassland and arable land, the main arable crops are sugar beet, potatoes, wheat and spring barley. It is a rare soil in the Netherlands.

Discussion

In most West European systems this soil is a pseudogley or surface-water gley. Such soils have 'strong mottling due to temporary waterlogging, above and in the B-horizon' (Dudal et al., 1966, p. 5); they are 'Dense soils in which the infiltrating water cannot enter or can enter the subsoil only very slowly' (Mückenhausen et al., 1977, p. 127); the same is stated by the French: 'The hydromorphism is temporarely and. . . it is caused by lack of infiltration of the rain water' (CPCS, 1967, p. 75). The difference in concept between the true gley and the pseudogley soils finds expression in the names of the major soil groups in the system of England and Wales: ground-water gley soils and surface-water gley soils respectively and in their definitions: presence or absence of a G or Cg horizon affected by free ground water (Avery, 1973).

Most surface-water gley soils have a slowly permeable subsoil, such as a buried palaeosol or a dense argillic horizon (Mückenhausen, 1973, p. 148). Such a subsoil is called in German: *Stauwassersohle, Staunässesohle* or *Staukörper,* and the more permeable upper layer: *Stauwasserleiter* or *Stauzone.* However, the opposite situation (fine pores over coarse pores, such as loess overlying coarse sand) may also be the cause of stagnating water (Gardner, 1968). The water relationships of this soil are even more complicated because the hydromorphism is not caused only by the latter reason, but also by ground water, for the water in the pervious substratum fluctuates into the solum. For this reason two possible classifications in the systems of our neighbouring countries are indicated on opposite page.

Both the legend of the Soil Map of the World and Soil Taxonomy do not discriminate between surface- and ground-water gley phenomena so the above-discussed difficulty does not exist in these systems.

Van den Broek & Van der Marel (1968/1969) discussed at length a soil sampled at the same site; they point out that the combination of hydromorphism and the high acidity not only results in clay illuviation from the upper part, but also in a strong weathering ('break-up of the clay minerals'; 'Soluvation: eluviation after solution of mineral components'; 'strong decomposition of the mineral soil particles', 'The designation of the profile as a podzol is justified when these evidences are used as criteria for classification'). It seems that these kind of soils closely resemble the sod-podzolic soils of the USSR (see also p. 162 and the literature cited there). They are quite different from the podzols of the glacial and periglacial sand plains of northwestern Europe (such as soil P1, p. 157 and P3, p. 165), but in the Netherlands hydromorphic soils developed in loess are rare.

The horizon nomenclature in the discussed systems is given in the following table:

Dutch	World	USA'75	Ger. (old)	Ger. (new)	E & W	French
A0	O	O	A0	O	L+F+H	A0
A1	Ah	A1	A1	Ah	Ah	A1
A2g	Eg	A2g	g1	Sw	Eg	A2g
Btg	Btg	Btg	g2	Sd	Btg	Btg
Dg	2Cg	IICg	DGo	IIGo	2Cg	IICG

The Germans have (and had) separate codes for pseudogley soils; but also the French, for the subscript g is only used for pseudogley phenomena ('normal' mottled horizons have to be coded BG or CG, compare p. 79).

The analytical data are characteristic for this acid hydromorphic soil. The texture is typical for a loess, having less than 20% of both clay and sand, the lowest sample shows a certain admixture from the underlying sand. The texture profile shows eluviation and illuviation. There is a high C-N ratio in the humic horizons and the whole solum is acid with very low values for the base saturation; the exchange capacity per 100 g of clay is least in the eluvial horizon.

Profile description

A0	+4-0 cm	Matted litter layer: very dark brown (7.5YR2/2) material; many well de-composed remnants of oak, beech and bracken; abrupt, smooth to wavy boundary.
A1	0-6/10 cm	Very dark grey brown (10YR3/2) silt loam; weak subangular blocky to crumb structure; friable, soft; clear, smooth to wavy boundary.
A2g	6/10-25/30 cm	Light grey (10YR7/1) silt loam; common, distinct, fine yellowish brown (10YR5/4) mottles; massive to weak medium platy breaking to weak very fine subangular blocky structure; gradual, irregular (glossic) transition to:
B2tg	25/30-70 cm	Strongly variegated silt loam; yellowish brown (10YR5/8) to reddish yellow (7.5YR6/8) colours predominate, next to light grey (10YR7/1), coatings on structural peds are pale brown (10YR6/3); many, prominent, coarse mott-les; weak, very coarse prisms breaking to medium angular blocky peds; firm, hard.
B3tg	70-110 cm	Transitional layer to the substratum (gravelly coarse sand), consisting of mixed coarse sandy loess.

Both solum and substratum are non-calcareous.

Analytical data

Horizon	Depth (cm)	Organic fraction				Particle-size distribution of mineral fraction (% of fine earth)			
		O.m. (%)	C(%)	N(%)	C/N	< 2	2-16	16-50	> 50 μm
A0	+4-0	39.2	20.4	0.97	21.0	·13	19	50	18
A1	0-6	5.3	2.4	0.09	26.7	10	14	55	20
A2g	10-25	0.7	n.d.	n.d.	n.d.	13	13	56	18
B2tg	40-50	0.6	n.d.	n.d.	n.d.	19	13	49	19
B3tg	81-90	0.4	n.d.	n.d.	n.d.	17	12	28	43

Hori-zon	Fe_2O_3 (%)	pH-KCl	Extractable cations (% of sum)					C.E.C. (meq./ 100 g)
			Na	K	Mg	Ca	H	
A0	0.99	3.2	n.d.	n.d.	n.d.	n.d.	n.d.	50.1
A1	1.08	3.6	1.0	0.9	1.1	2.8	94.3	10.2
A2g	1.01	4.0	2.1	1.0	1.9	5.2	89.8	3.2
B2tg	3.17	3.6	1.0	1.5	1.9	4.6	91.0	8.8
B3tg	2.37	3.6	1.3	1.5	3.5	7.8	85.8	8.6

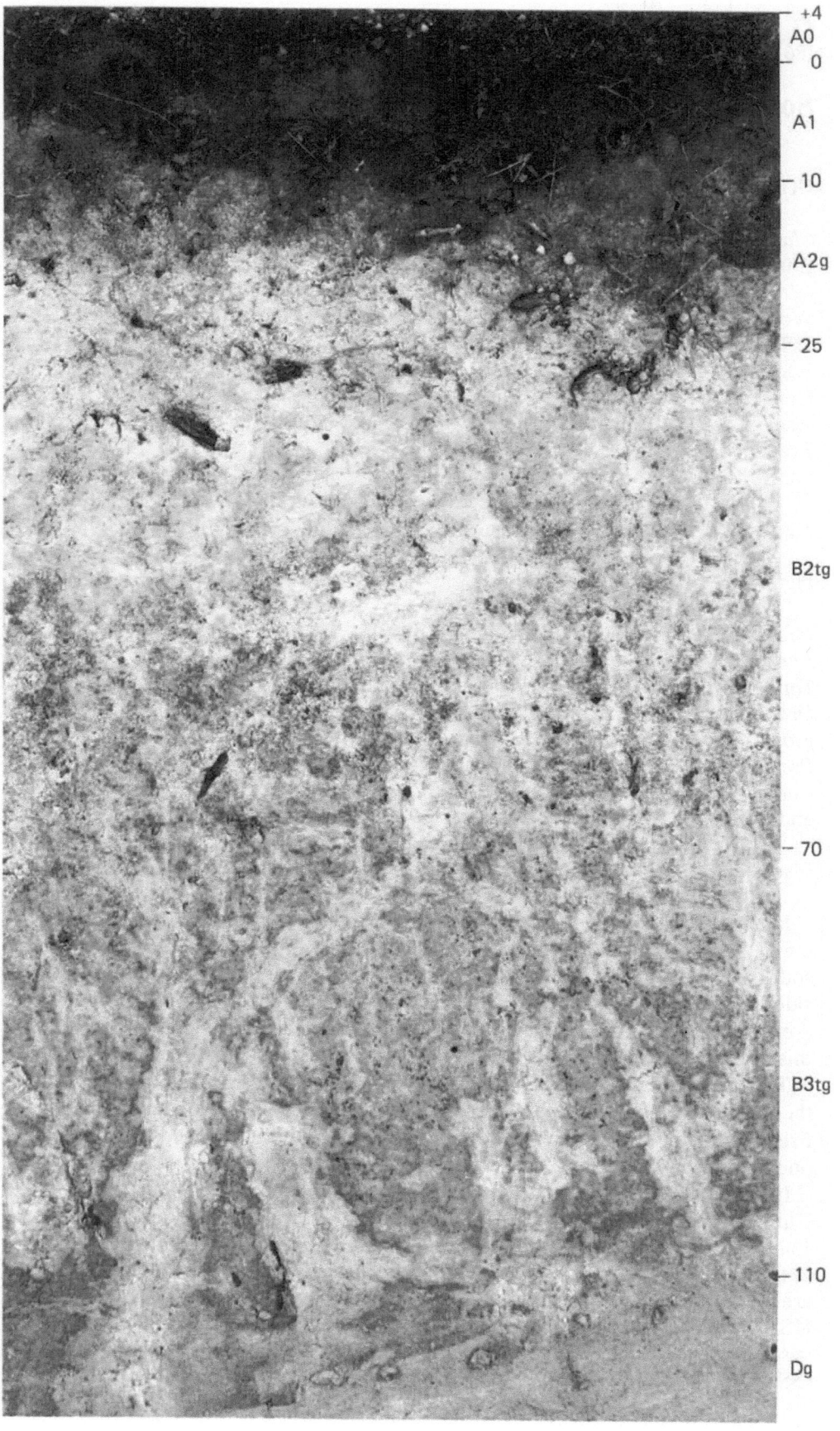

Soils of the loess district: L3

General data

Classification

Eur.	Rendzina (p. 4), very small inclusions in the association of Gray-brown podzolic soils in the south of the Netherlands
World	Rendzina (p. 34), medium textured, rolling (p. 5)
USA '38	Rendzina soil (p. 1001 and 1106-1110)
USA '75	Loamy-skeletal, carbonatic, mesic Lithic Rendoll (p. 294)
Ger.	Mullrendzina (p. 65)
Eng. & Wales	Humic rendzina
France	Rendzine modale (p. 36)
Neth. '50	Not mentioned in the list of mapping units nor in the book; included in the mapping unit 55
Neth. '61	Not mentioned specifically; included in the mapping units 148 and 149
Neth. '66	'Krijt' earth soil (p. 155 and 194); mapping unit KD

Location. Limburg Province, about 15 km east of Maastricht (see Fig. 1).
Parent material. Cretaceous chalk.
Topography and elevation. Sloping and strongly sloping; ca. 90 m above sea level.
Drainage and ground water. No external drainage, runoff only during showers; no ground-water influence.
Present land use. Grassland.
Range in land use. These soils are rare in the Netherlands. They occur on slopes and usually carry deciduous forest, locally grassland.

Discussion

In the Netherlands soils developed on pre-Quaternary rocks are rare, because of the extensive area of Pleistocene and Holocene deposits. Only in Limburg, in the south of the country, does the Cretaceous chalk crop out and then only on valley sides as the intervening plateaux are loess-covered (Fig. 2 and 34). The rendzinas are geographically associated with the loess soils and are considered with them in this section.

The term rendzina originally is Polish, 'the name is used by Polish farmers. . . the spelling in the Polish language is 'rędzina' (because ę is pronounced like the French 'in'), the term is derived from the word 'rzędzić' = to drone, after the noise one hears when ploughing' (Miklazewski, 1924).

In virtually all systems this term has been adopted (see above and Kubiëna, 1953, p. 212; Ehwald et al., 1966, p. 45 and 46; Němeček et al., 1967, p. 134-139; Rozov & Ivanova, 1967; Kuźnicki et al., 1974, p. 15; FitzPatrick, 1971, p. 221), even in Soil Taxonomy with its new names the first four letters of the Rendolls are derived from the old name. However, when comparing the old and the new USA systems it appears that none of the soil series which were rendzinas in the old system of the USA (USDA, 1938, p. 1106-1110) today are classified as Rendolls (SSS, 1972). Most of them are either other Mollisols or Vertisols.

Some systems of soil horizon nomenclature use the designation A1 for dark surface horizons, such systems use the suffix p 'to indicate disturbance by plowing or pasturing' (SSS, 1975, p. 462), or according to the French system for a 'ploughed (or disturbed) horizon' (CPCS, 1967, p. 13); the omission of the suffix p points to the virgin character of the A1 horizon. Other systems use the suffix h instead of the number 1 (the FAO, the German and the British systems) thus indicating positively that 'there has been no disturbance or mixing from ploughing, pasturing or other activities of man (h and p suffixes are thus mutually exclusive)' (FAO, 1974, p. 22). The soil is used as permanent grassland and probably has never been ploughed. In the Netherlands topsoils of such soils are not coded Ap (cf. e.g. soil M3, p. 73 and soil F1, p. 93), but according to the letter-of-the-law of Soil Taxonomy and the FAO the suffix p must be used, because these soils are pastured! In the Netherlands the suffix p is used exclusively for horizons and layers which are actually mixed (cf. soil RB2, p. 121, soil RB4, p. 129 and soil RB5 p. 133), like in England and Wales: 'Ap – Surface mineral horizon evidently mixed by cultivation (Hodgson, 1974, p. 7).

The parent rock is coded Cr, an identical code is found only in the system of England and Wales: 'Cr – A weakly consolidated, little-altered substratum . . . Such rocks as . . . chalk . . . are included' (Hodgson, 1974, p. 79 and 80). This system and also the system of the ISSS, as published by the FAO (1974, p. 20-23), subdivide substrata into three classes: unconsolidated, weakly consolidated and hard bedrock, designated Cu, Cr and R in the British system and C, Cm and R in the FAO system. Soil Taxonomy and the French system have two possibilities: C and R. All these systems use the prefix 2 or II to indicate a lithological discontinuity, which is not present in this soil: the mineral particles in the upper horizons are supposed to be the weathering residu from the chalk. In the German system it is not possible to indicate whether a substratum is consolidated or unconsolidated.

The analytical data are representative for such soils: a high organic-matter content; a low C-N ratio; the soil is nearly saturated with Ca-ions; the CEC is high but this is mainly due to the high organic-matter content; the parent rock has nearly 90% calcium carbonate (analysed CO_2, calculated as $CaCO_3$).

Profile description

A11	0-5 cm	Very dark grey (2.5Y3/1) silt loam; few, fine pieces of chalk; weak to moderate fine granular to fine crumb; friable, slightly hard; abrupt, smooth boundary.
A12	5-15 cm	Very dark grey (2.5Y3/1.5) loam to silt loam; more, coarser pieces of chalk; moderate fine to medium granular structure; friable, slightly hard; abrupt smooth boundary.
A13	15-30 cm	Transitional horizon to bedrock: dark grey brown (10YR3/2) sandy clay loam; many, medium and coarse pieces of chalk, some flints; moderate medium granular to fine subangular blocky structure; clear to gradual, smooth boundary.
Cr	30-120 cm plus	The soil grades into the bedrock; pale yellow (5Y8/3) chalk; firm to very firm, very hard.

Analytical data

Hori-zon	Depth (cm)	Organic fraction				Particle-size distribution of mineral fraction (% of fine earth)					
		O.m. (%)	C(%)	N(%)	C/N	< 2	2-16	16-50	50-105	105-150	> 150 µm
A11	0-5	9.7	5.0	0.36	13.9	18	10	46	3	4.5	19
A12	5-12	5.4	3.0	0.24	12.5	20	9	41	3.5	5	21
A13	20-30	3.2	n.d.	n.d.	n.d.	28	12	10	5	9	36
Cr	50-60	0.3	n.d.	n.d.	n.d.	n.d.	n.d.	n.d.	n.d.	n.d.	n.d.

Hori-zon	CaCO₃ (%)	pH-KCl	Extractable cation (% of sum)					C.E.C. (meq./100 g)
			Na	K	Mg	Ca	H	
A11	24.3	7.1	1.2	2.4	3.3	90.9	2.1	33.1
A12	27.9	7.4	n.d.	n.d.	n.d.	n.d.	n.d.	n.d.
A13	37.0	7.3	n.d.	n.d.	n.d.	n.d.	n.d.	n.d.
Cr	86.6	8.7	n.d.	n.d.	n.d.	n.d.	n.d.	n.d.

0

A11

5

A12

15

A13

30

Cr

Appendix 1: Terminology and spelling

For a proper understanding of any subject in whatever language it is essential to understand the terminology used. Endless discussions, especially between pedologists, are caused primarily by misunderstanding the terminology, even when speaking the same language. In fact, one of the intentions of this book (expressed in the Discussion sections in the last chapter) is to show that it is only possible to review different opinions about soils, their genesis and classification, horizon nomenclature, and so on, if the terminology is reasonably well defined.

Many terms in this book do not need precise definition (plough layer, shallow soil, waterlogged, raised bog, gleyed). Other terms could have been explained in a glossary, but this has been omitted; most of those terms must be clear from the context, but some will be mentioned below.

Soil terminology. In most places in the text it is made clear from the literature cited in what sense a soil term is being used; if no references are given, it may be assumed that Soil Taxonomy terminology is used. This is also true for the soil maps (Figures 5, 12, 16, 19, 23 and 28). This system is widely known and discussed, and unlike the other systems it is detailed enough to classify all mapping units: the classes are clearly defined in terms of morphological properties.

Botanical nomenclature. The Dutch edition of Polunin's *Flowers of Europe* (1969) was used for the scientific botanical names.

Sea level. This expression is used to indicate the Dutch reference level, called *Normaal Amsterdams Peil* (NAP): it is a level fixed on a 22 m long pole rammed below the Dam in Amsterdam and lying about 9 cm above mean sea level in Den Helder. It is the same level as the German Normalnull (NN), it is 2.42 m higher than the Belgian Oostends Peil (OP), and 23 cm higher than the British Ordnance Datum (OD).

Some polder terminology. A polder is: 'A piece of land surrounded by dikes (embankments) in which the water level in the ditches is artificially maintained' (Pons & Van der Molen, 1973). Not all Dutch polders are reclaimed by enclosing coastal marshland (p. 21), or by draining shallow lakes (p. 33) or shallow parts of the former Zuyder Zee (p. 23). This term is also used for areas of land that have not been won from the sea, but where artificial control of water levels in already reclaimed land, e.g. peat lands or the older lands in the marine district (p. 19), has allowed the land to be better drained.

The word 'polder' is hardly used in English Soil Survey Bulletins. Green, who surveyed Romney Marsh in Southern England (1968) obviously prefers the term 'enclosure', both for the action of enclosing a salting (coastal foreland subject to tidal flooding) and for the enclosed area (a polder); he also uses the word 'innings' ('lands recovered from the sea', *Chambers Twentieth Century Dictionary,* new edition 1972, reprinted 1974).

The word *dijk* (Dutch) is generally translated by 'dike' or 'dyke'; this might be confusing for British readers because its meaning (apart from the geological one) may also be 'ditch' (larger open drain in polders); in English coastal areas the word 'wall' (sea wall, innings wall) is usually used for these sea defences, which nearly everywhere are earthen embankments rising 4-7 m above sea level (depending on the average height of the high tide, exposure to prevailing storms, presence of forelands, and the like). Outer sea walls are called *wakers* (Dutch) = watchers, inner (older) sea walls *slapers* (Dutch) = sleepers (cf. Fig. 5).

The embankments built to protect the Holocene floodplain of the rivers from flooding, are called 'artificial levees' in this book. (The natural, wide, slightly elevated banks built up on both sides of a river are called 'natural levees'). The high artificial levee that must stem winter floods is called *winterdijk* (Dutch) = winter dike, and the low artificial levee that must protect the forelands from the rare summer floods is called *zomerdijk* (Dutch) = summer dike (cf. Fig. 12 and 13).

Bedijken or *indijken* (Dutch), 'to enclose an area with sea walls' is often translated in Dutch publications by 'to embank', 'to empolder' might be a better translation, or even 'to dike'. 'To enclose' might cause confusion; the enclosure of open fields is quite a different matter compared with enclosing coastal marshes. 'To reclaim coastal marsh land' is the phrase used in English writing; however, this might also be confusing because this verb is also used for converting heathland into pasture or arable land. In the large Zuyder Zee polders it takes some years after enclosure before the whole polder is completely reclaimed.

Spelling. Quotations from literature other than English or American are translated, the original quotations are assembled in Appendix 3. Quotations are always printed in the original spellings: sulphuric (Avery, 1973) and sulfuric (SSS, 1975, p. 47); Humic-Glei soil (Thorp & Smith, 1949, p. 118) and humic gley soil (Avery, 1973); levee (*Chambers Dictionary*) and calcareous river levée soils (mapping unit 25 on Edelman's map accompanying his book from 1950).

Appendix 2: Analytical methods

Opposite each colour plate is a profile description with some analytical data. The latter include particle-size distribution, with a detailed distribution of the sand fraction of some horizons for coarse-textured soils. Cation-exchange capacity (CEC) and exchangeable cations have been analysed for only some horizons of medium- and fine-textured soils.

The kind of data and the analytical methods are discussed below.

pH was measured in a 1:5 suspension of soil in 1 N potassium chloride solution.

CaCO3 content was determined by measuring the volume of carbon dioxide released when the soil was treated with 25% hydrochloric acid, and calculated to calcium carbonate equivalent.

Organic matter was determined by:

a. the loss-on-ignition (of weight) at 600°C or 950°C. Corrections are made for the loss of water (9.6% of the clay content, in both cases) and for the loss of CO_2 (0 and 44% respectively of the carbonate content).

b. wet oxidation with $KMnO_4$ (Istcherikov method). As the oxidation is incomplete an empirically determined correction factor, varying between 1.3 and 1.5 was applied to the calculated total organic carbon. The C-content of the organic matter was taken as 58%.

Organic carbon was gravimetrically determined from the amount of carbon dioxide released on dry combustion at 950°C. In calcareous samples a correction is made for the amount of carbon dioxide released from the calcium carbonate.

Particle-size distribution was determined by the pipette method after pretreatment of the samples with 30% hydrogen peroxide and 0.2 N hydrochloric acid with sodium pyrophosphate and soda. After the < 16 μm separate and the clay separate were taken, the sand preparate was removed by wet sieving. The subfractions of the sand separate were then collected by dry sieving. The initial weight of fine earth was corrected for organic matter, carbonate content and the 16-50 μm separate obtained by difference.

Cation-exchange capacity (CEC) was obtained by percolation of the soil with 0.5 N calcium acetate solution of pH 8.2 and 6.5, respectively, and measurement of the calcium ions released upon leaching with 0.1 N sodium chloride. The results are corrected, if necessary, for dissolved calcium carbonate by titration of the percolate with hydrochloric acid.

Exchangeable cations were determined in a 1 N ammonium nitrate leachate, for sodium and potassium, and in a 1 N sodium chloride leachate, for calcium and magnesium. The exchangeable calcium is corrected for dissolved calcium from calcium carbonate and gypsum. Exchangeable hydrogen was determined by titration of the 0.5 N calcium acetate solution (pH 8.2) used in the CEC determination.

Potassium-fixation was measured by shaking 2.5 g of the sample with 10 ml potassium chloride solution containing 2.5 mg of K_2O. The sample was then leached using 0.25 N magnesium acetate and the potassium concentration measured in the leachate. Exchangeable potassium was determined by the same method, but without potassium chloride, to allow for potassium release by the sample. The potassium retained by the sample, i.e. the potassium-fixation, has been calculated as the percentage of the potassium added.

Iron was extracted by boiling 10 g of soil for half an hour in a 10% hydrochloric acid solution and the amount determined by titration and calculated as Fe_2O_3.

Total Ca and SO₄ were determined in soils containing pyrites to establish their potential acidity. 15 ml of a 3:1 mixture of 65% nitric acid and 36% hydrochloric acid was added to 5 g of a sample and boiled. Ca and SO_4 were measured in the filtrate.

Total nitrogen was determined by a modified Kjeldahl method. 2-5 g of soil was digested with 15 ml of a concentrated sulphuric acid – phenol mixture, 6 g of catalyst (with Na_2SO_4, $CuSO_4$, $5H_2O$ and Se) and 20 ml of concentrated sulphuric acid. The NH_3 was distilled into H_3BO_3 and titrated with standard sulphuric acid.

A-figure is the moisture content in g of water per 100 g of ovendry soil (105°C).

194

Appendix 3: Quotations

p. 63: 'Dieses Profil ist bereits bis zu einer Tiefe von 70-80 cm belüftet' (Hugenroth, 1971, p. 38).

p. 71: '. . . on parlera, suivant le cas, de sol peu évolué d'apport marin, ou de sol hydromorphe minéral à gley. . .' (Servant, 1973, p. 40).

p. 74: '. . . Körngröszenzusammensetzung der Marschhorizonte bzw. -schichten. . .' (Mückenhausen et al., 1977, p. 158).

p. 83: '. . ., on peut diviser les sols d'une même sous-groupe en tenant compte du matériau originel' (CPCS, 1967, p. 9).

p. 91: '. . . plus que des traces de matière organique dans les 20 centimètres supérieurs. . .' (CPCS, 1967, p. 21). '. . . à certaines périodes à moins de 1 m' (CPCS, 1967, p. 25). 'Il n'y a jamais, dans ces sols, d'horizons A2, B, ni même (B)' (CPCS, 1967, p. 21).

p. 103: '– plus de 30% sur au moins 40 cm si la matière minérale est argileuse. – plus de 20% si la matière minérale est sableuse' (CPCS, 1967, p. 73).

p. 111: '. . . nepromyvného -periodicky promyvného vodního režimu' (Němeček, 1967, p. 108).

p. 126: 'Aber auch die durch Tiefpflügen gewendeten Böden wollen wir zu dem Rigosoltyp stellen'. (Mückenhausen, et al. 1977, p. 140).

p. 135: '. . ., aannemende, dat een bunder lands alle drie jaren met 80 voeren heideplaggen bemest werd en dat deze, na vergaan te zijn, 40 teerling ellen zwarten grond achterlieten, dan zoude de grond daarmede telkens 4 strepen worden opge-hoogd en eene verhooging van ééne el, derhalve 750 jaren vereischen' (Staring, 1856, p. 12).

An interesting detail in this quotation is, that Staring being a purist, used the obsolete words *bunder* (hectare), *ellen* (metres) and *strepen* (millimetres). When the Netherlands adopted the metric system (by law in 1819, but in practice gradually in the course of the nineteenth century) it was allowed, next to the artificial terms like millimetre, to continue to use the obsolete terms in the new sense by adding the adjective *Nederlandsche*. For the word *kubieke* (cubic) Staring used an old word, meaning die (*teerling*).

p. 143: '. . . ne que des traces' (CPCS, 1967, p. 17). '. . . plus que des traces de matière organique' (CPCS, 1967, p. 21).

p. 147: '. . . durch jahrzehnte – bzw. jahrhundertelange, intensive Gartenkultur, . . .' (Mückenhausen et al., 1977, p. 139).

p. 155: '. . . sind die Bodenbildungen auszerhalb des Grundwasserbereiches ver-einigt; . . .' (Mückenhausen et al., 1977, p. 51). '. . ., der im oberen Profilteil einen Podsol und im unteren einen Gley darstellt' (Mückenhausen et al., 1977, p. 111). 'Au-dessus de Bh, il y a un Go ou Gr de gley' (CPCS, 1967, p. 54).

p. 159: 'Podsol mit einer Plaggenauflage bis zu 40 cm Mächtigkeit' (Mückenhausen et al., 1977, p. 111).

p. 163: 'Der B1-horizont enthält eine Illuviation von überwiegend Brauneisen und organischer Substanz' (Mückenhausen et al., 1977, p. 110).

p. 167: 'ДЕРНОВО-СЛАБОПОДЗОЛИСТАЯ ПЕСЧАНАЯ ПОЧВА'

(Kauritsjeva & Gromyko, 1974, p. 42). '. . . Mineralbodenhorizont, der i.a. nahezu gleichmäszig durch Sauerhumusstoffe gefärbt ist und meist gebleichte Quarzkörnchen enthält' (Arbeitsgemeinschaft Bodenkunde, 1971, p. 29). 'durch Verwitterung verbraunter und verlehmter Horizont zwischen dem A- und C-Horizont ohne oder ohne nennenswerte Illuviation, . . .' (Arbeitsgemeinschaft Bodenkunde, 1971, p. 29).

p. 182: 'Dichtgelagerte Böden, in denen das Sickerwasser nicht oder nur sehr langsam in den Untergrund abziehen kann' (Mückenhausen et al., 1977, p. 127). 'L'hydromorphie est temporaire et. . . elle est due au manque d'infiltration des eaux pluviales' (CPCS, 1967, p. 75).

p. 186: '. . . le nom est employé par les paysans polonais. . . on écrit en polonais rędzina (par ę ce qu'on prononce 'in')' '. . . mot 'rzędzić' – d'après les sons qu'on entend en le labourant' (Miklazewski, 1924).

p. 187 Horizon labouré (ou perturbé)' (CPCS, 1967, p. 13).

Appendix 4: Sources of text figures and tables

Fig. 2: This map is derived from the Soil Map of the Netherlands, scale 1:50 000 (as far as surveyed in 1975); from the Soil Map of the Netherlands, scale 1:200 000 (Stichting voor Bodemkartering, 1961) and from several unpublished maps at different scales, surveyed by the Dutch Soil Survey Institute; the distribution of the loess in Belgium and in the German Federal Republic is taken from the Soil Association Map of Belgium, scale 1:500 000 (Atlas van België, 1950-1972, blad 11 B, 1970) and from the Soil Map of Nordrhein-Westfalen, scale 1:500 000 (Maas & Mückenhausen, 1970).

Fig. 3: Contourlines from the Ordnance Survey of the Netherlands, scale 1:25 000, Topographical Service, Delft.

Fig. 4: Photo KLM-Aerocarto, 2-5 1966, Topographical Service, Delft. File nos. 43 VII 223 and 225.

Fig. 5: Simplified segment from sheet 43 West of the Soil Map of the Netherlands, scale 1:50 000.

Fig. 6: Data between 1200 and 1948 after Van Veen (1948); supplemented by the acreage embanked since 1948.

Fig. 7: After Figure 7 in the memoir to sheet 43 West and after Figure 7 in the memoir to sheet 43 East of the Soil Map of the Netherlands, scale 1:50 000.

Fig. 8: Simplified and reduced segments from sheets 10 and 15 of the Soil Map of the Netherlands, scale 1:50 000.

Fig. 9: Derived from Figure 8.

Fig. 10: Photo KLM-Aerocarto, 10-4 1971, Topographical Service, Delft. File nos. 20 O VI 63 and 65.

Fig. 11: Photo KLM-Aerocarto, 8-4 1974, Topographical Service, Delft, File nos. 39 VI 192 and 194, 39 VII 225, 227 and 228.

Fig. 12: Simplified segment, partly from sheet 39 West and East of the Soil Map of the Netherlands, scale 1:50 000 and partly from the Geomorphological Map of the Forelands, scale 1:50 000 (in: De Soet et al., 1975).

Fig. 13: Photo APIS (Allied Photo Intelligence Service?), I CDN Sqr, F/L Oglive, 21-2 1945, 12h40, 28 000 ft, F36″.

Fig. 14: After Figure 20 in the memoir to the sheets 39 West and East and after Figure 17 in the memoir to the sheets 40 West and East of the Soil Map of the Netherlands, scale 1:50 000; the southern part is simplified from the sheets 45 West and East and 46 West and East of the same soil map.

Fig. 15: Photo KLM-Aerocarto, 14-4 1967, Topographical Service, Delft. File nos. 31 II 110, 112 and 113.

Fig. 16: Simplified segment from sheet 31 West of the Soil Map of the Netherlands, scale 1:50 000.

Fig. 17: The innings data after Figure 11 in the memoir to sheet 31 West and the physiographical data after Figure 12 in the memoir to sheet 31 West and Figure 16 in the memoir to sheet 31 East of the Soil Map of the Netherlands, scale 1:50 000.

Fig. 18: Photo Hansa Luftbild, 13-7 1972, Topographical Service, Delft. File nos. 18 V 216, 218 and 219.

Fig. 19: Simplified segment from sheet 18 of the Soil Map of the Netherlands, scale 1:50 000.

Fig. 20: Derived from Figure 19.

Fig. 21: Much simplified portion of sheet 1 of the Soil Map of the Netherlands, scale 1:200 000.

Fig. 22: Photo KLM-Aerocarto, 25-4 1973, Topographical Service, Delft. File nos. 27 V 137, 27 VI 63 and 64.

Fig. 23: Simplified fragment from Appendix 2B (Steur & De Bakker, 1969).

Fig. 24: Simplified fragment from Appendix 3 (Ten Houte de Lange, 1977).

Fig. 25: According to Pape (1970), Figure 2.

Fig. 26: Photo KLM-Aerocarto, 22-4 1976, Topographical Service, Delft. File nos. 37 W, 116.

Fig. 27: Photo KLM-Aerocarto, 17-5 1973, Topographical Service, Delft. File nos. 27 VII 155, 156 and 158.

Fig. 28: Simplified fragment of sheet 27 East of the Soil Map of the Netherlands, scale 1:50 000, the eastern third part of this figure after unpublished data.

Fig. 29: After the Ordnance Survey of 1882.

Fig. 30: Mainly after the yearly Reports of the Dutch Ministry of Agriculture.

Fig. 31: The Pleistocene part after Figure 1 of Schelling & Marsman (1973), the Holocene part after the Soil Map of the Netherlands, scale 1:200 000, sheet 3.

Fig. 32: Photo KLM-Aerocarto, 23-2 1975, Topographical Service, Delft. File nos. 62 IV 75 and 76.

Fig. 33: Unpublished field data from Mr. J. A. M. ten Cate.

Fig. 34: Partly after the Soil Map of the Netherlands, scale 1:50 000 (sheets 58, 59 and 60); partly after the Soil Map of the Netherlands, scale 1:200 000 (sheet 9); partly after the Soil Map of Nordrhein-Westfalen, scale 1:50 000 (sheets L4902 and L5000); partly after the Soil Map of Nordrhein-Westfalen, scale 1:500 000 (Maas & Mückenhausen, 1970); partly after the Soil Associations of Belgium (Atlas, 1950-1972, sheet 11B).

Fig. 35: Figure 11 in the memoir to sheets 39 West and East of the Soil Map of the Netherlands, scale 1:50 000. Photo R24-5, Soil Survey Institute.

Fig. 36: Photo Koninklijke Nederlandsche Heidemaatschappij.

Fig. 37: Photo Koninklijke Nederlandsche Heidemaatschappij.

Fig. 38: After De Bakker & Marsman (1978).

All colour plates and the figures opposite the plates of the soils RB 4 and RB 5 are from De Bakker & Edelman-Vlam (1976).

Table 1: Files from the Netherlands Soil Survey Institute.

Table 2: Data from the Royal Netherlands Meteorological Institute.

Table 3: From Visser (1958, p. 146).

Table 4: Original.

Table 5: After unpublished data from the *Rijkswaterstaat* (Netherlands Water Authority).

198

Acknowledgments

In an overall view of soils and landscapes of the Netherlands, like this book, there are many colleagues and friends who have contributed, advised and given moral support to the author.

Some of them I want to mention specifically. Firstly Mr. W. J. M. van der Voort who did the laborious job of making and preparing the monoliths; further the photographers at our Institute, Mr. M. C. Nater and Mr. C. T. van der Schouw, who did not object to having the dirty monoliths in their clean darkroom.

Thanks are also tendered to the following colleagues: Dr. A. Breeuwsma (author of Appendix 2); Ing. B. A. Marsman, Ir. J. C. Pape, Ir. P. van der Sluijs and Drs. J. A. J. Vervloet (for the stimulating discussions); Mr. G. Staal (for advising about text figures, especially about the aerial photographs); Mr. H. C. Bos, Mr. W. Bos and Mr. C. P. van der Spek (for their excellent drawings); Dr. J. Schelling, Ir. G. G. L. Steur and Mr. J. W. Zwolschen (for reading and commenting on earlier concepts).

The English text was read partly by Dr. E. M. Bridges (senior lecturer in Geography at University College of Swansea) and Mrs. G. Bridges (thanks for both scientific and editorial discussions) and partly by Mrs. J. Boenisch-Burrough. There must still be wrong English in this book, because of 'improvements' by the author.

Literature

Arbeitsgemeinschaft Bodenkunde, 1971. Kartieranleitung. Anleitung und Richtlinien zur Herstellung der Bodenkarte 1:25 000. Bundesanstalt f. Bodenforschung und Geologische Landesämter der BRD, Hannover.

Atlas, 1950-1972: Atlas van België. Koninkrijk België, Nationaal Comité voor Geografie, Commissie voor de atlas. Brussel.

Atlas, 1963-1977: Atlas van Nederland (Atlas of the Netherlands, 100 sheets. Dutch and English. Compiled by the Foundation for the Scientific Atlas of the Netherlands). Staatsdrukkerij- en Uitgeverijbedrijf, 's-Gravenhage.

Avery, B. W., 1973. Soil classification in the Soil Survey of England and Wales. J. Soil. Sci. 24: 324-338.

Bennema, J., J. Schelling & J. S. Veenenbos, 1953, 'Great Soil Groups' in Nederland (with summary: Great soil groups in the Netherlands). Boor en Spade 6:41-51. Veenman, Wageningen.

Booij, A. H., 1959. Drentse dalgronden, uniforme gronden? (with summary: Reclaimed peat moor soils of Drente. Uniform soils?). Boor en Spade 10: 97-105. Veenman, Wageningen.

Breteler, H. G. M., 1958. Kleefaarde (with summary: Sticking earth). Boor en Spade 9: 62-70. Veenman, Wageningen.

Breteler, H. G. M. & J. M. M. van den Broek, 1968. Graften in Zuid-Limburg (with summary: Escarpments in South Limburg). Boor en Spade 16: 119-130. Veenman, Wageningen.

Bridges, E. M., 1973. Some characteristics of Alluvial Soils in the Trent Valley, England. In: Pseudogley & Gley, Trans. Comm. V and VI of the ISSS: 247-253. Verlag Chemie GmbH, Weinheim/Bergstr.

Brümmer, G., 1968. Untersuchungen zur Genese der Marschen. Thesis Kiel.

Casparie, W. A., 1972. Bog development in southeastern Drenthe (The Netherlands). Vegetatio 25: 1-4. Thesis Groningen.

Clayton, J. S., W. A. Ehrlich, D. B. Cann, J. H. Day & I. B. Marshall, 1977. Soils of Canada. A cooperative project of The Canada Soil Survey Committee and The Soil Research Institute. Research Branch Can. Dep. Agric., Ottawa.

Conry, M. J., 1969. Plaggensoils in Ireland. Pédologie, Ghent 19: 321-329.

Conry, M. J., 1971. Irish Plaggensoils – their distribution, origin, and properties. J. Soil. Sci. 22: 401-416.

Conry, M. J., 1974. Plaggensoils, a review of man-made raised soils. Soils Fertil. 37: 319-326.

Conry, M. J. & J. J. Diamond, 1971. Proposed classification of plaggensoils. Pédologie, Ghent 21: 152-161.

CPCS, 1967. Classification des sols. Commission de Pédologie et de cartographie des sols.

CSSC (Canada Soil Survey Committee), 1970. The system of soil classification for Canada. Canada Dep. Agric., Queen's Printer, Ottawa.

De Bakker, H., 1965. Tonverlagerung in Fluszablagerungen verschiedener Art. Mitt. dt. Bodenk. Ges. 4: 123-128. Göttingen.

De Bakker, H., 1970. Purposes of soil classification. Geoderma 4: 195-208.

De Bakker, H., 1971. The distinction between Haplaquents and Haplaquepts in the Dutch marine and fluviatile sediments. Geoderma 5: 169-178.

De Bakker, H. & J. Schelling, 1966 (unchanged reprints 1974, 1976). Systeem van bodemclassificatie voor Nederland, de hogere niveaus (with 30 pages summary: A system of soil classification for the Netherlands, the higher levels). Pudoc, Wageningen.

De Bakker, H. & A. W. Edelman-Vlam, 1976. De Nederlandse bodem in kleur (no summary). Stichting voor Bodemkartering/Pudoc, Wageningen.

De Bakker, H. & B. A. Marsman, 1978. Kruinige percelen (with summary: Domed fields). Boor en Spade 20 (in prep.). Veenman, Wageningen.

De Soet, F. (Ed.), 1976. De waarden van de uiterwaarden, een milieukartering en -waardering van de uiterwaarden van IJssel, Rijn, Waal en Maas (with summary: A study of the forelands). Pudoc, Wageningen.

Domhof, J., 1953. Strooiselwinning voor potstallen in verband met de profielbouw van de heide- en oude bouwlandgronden (with summary: Procurement of litter for built-up heath-sod litter manure in connection with the profile of heath-soils and old arable land). Boor en Spade 6: 192-203. Veenman, Wageningen.

Dost, H. (Ed.), 1973. Acid sulphate soils. Proc. Int. Symp. I, Introductory papers and bibliography, II, Research papers. Publ. 18.1 and 18.2, ILRI, Wageningen.

Duchaufour, Ph., 1965. Précis de pédologie. Masson, Paris.

Dudal, R., R. Tavernier & D. Osmond, 1966. Soil Map of Europe, 1:2 500 000, Explanatory Text. FAO, Rome.

Dijkerman, J. C., 1965. Properties and genesis of textural subsoil lamellae. Thesis Cornell University.

Edelman, C. H., 1950. Soils of the Netherlands. North Holland, Amsterdam.

Edelman, C. H., 1963. Bospodzolen en heidepodzolen (with summary: Forest podzols and heath podzols). Boor en Spade 13: 51-60. Veenman, Wageningen.

Ehwald, E., I. Lieberoth & W. Schwanecke, 1966. Zur Systematik der Böden der Deutschen Demokratischen Republik besonders im Hinblick auf die Bodenkartierung. Sitzungsberichte dt. Akad. Landw. Wiss. Berlin. Bd XV, Heft 18.

Eisma, D., 1968. Composition, origin and distribution of Dutch coastal sands between Hoek van Holland and the island Vlieland. Neth. J. Sea Res. 4: 123-267. Also thesis Leiden.

FAO, 1974. FAO-Unesco Soil Map of the World, 1:5 000 000, Volume I, Legend. Unesco, Paris.

FitzPatrick, E. A., 1971. Pedology, a systematic approach to soil science. Oliver & Boyd, Edinburgh.

Fowlkes, Th., C. G. Morgan, J. A. Herren, D. D. Mason & L. A. Davidson, 1956. Soil survey of Tunica county, Mississippi. Series 1942, No. 14, U.S. Dep. Agric. and Miss. agric. Expt. Station.

Gardner, W. H., 1968. How water moves in the soil. Crops Soils Mag., Nov. 1968, Madison.

Gerasimov, I. P. & S. V. Zonn, 1971. Podzol and gley; lessivé, pseudogley, and pseudopodzol (priority of genetic concepts). Soviet Soil Sci. 3: 496-506 (Translated from Pochvovedeniye, 8 (1971): 118-129).

Gibbs, H. S., 1960. Introduction and general review. In: The impact of man on soils – A symposium: 13-17. Proc. N.Z. Soc. Soil Sci. 4.

Goetz, D., 1970. Bänderparabraunerden aus jungpleistozänen Sanden im Raum Berlin. Thesis Institut für Bodenkunde der Technischen Universität Berlin.

Green, R. D., 1968, Soils of Romney Marsh. Soil Survey of Great Britain. England and Wales. Bulletin No. 4, Harpenden.

Hitchcock, S. W., 1972. Can we save our salt marshes? Natn. geogrl. Mag. 141: 728-765. Washington, D.C.

Hodgson, J. M., 1967. Soils of the West Sussex Coastal Plain. Soil Survey of Great Britain. England and Wales. Bulletin No. 3, Harpenden.

Hodgson, J. M. (Ed.), 1974. Soil Survey Field Handbook, Describing and sampling soil profiles. Soil Survey, Techn. Monograph No. 5. Rothamsted Exp. Station, Harpenden.

Hodgson, J. M., J. A. Catt and A. H. Weir, 1967. The origin and development of clay-with-flints and associated soil horizons on the South Downs. J. Soil Sci. 18: 85-102.

Hoeksema, K. J., 1953. De natuurlijke homogenisatie van het bodemprofiel in Nederland (with summary: The natural homogenization of the soil profile in the Netherlands). Boor en Spade 6: 24-30. Veenman, Wageningen.

Hollstein, W., 1963. Bodenkarte der Bundesrepublik Deutschland, in Maszstab 1:1 000 000. Bundesanstalt für Bodenforschung, Hannover.

Hugenroth, P. (Ed.), 1971. Landschaften und Böden in der Bundesrepublik Deutschland. Exkursionsführer zur Tagung der Kommissionen V und VI der Int. bodenk. Ges. in Stuttgart-Hohenheim, Exkursion C, Mitt. der D. Bodenk. Ges. Göttingen.

ISSS (International Society of Soil Science), 1967. Proposal for a uniform system of soil horizon designations (first draft, Sept. 1967). In: ISSS Bulletin No. 31, 4-7. Amsterdam.

Jamagne, M., 1967. Bases et techniques d'une cartographie des sols. Annls Inst. natn. Rech. agron., Sér. A. Annls agron. 18 (No. hors-serie).

Jenny, H., 1941. Factors of soil formation, a system of quantative pedology. McGraw-Hill, New York/London.

Jenny, H., 1961. Derivation of state factor equations of soils and ecosystems. Proc. Soil Sci. Soc. Am. 25: 385-388.

Jongerius, A., 1970. Some morphological aspects of regrouping phenomena in Dutch soils. Geoderma 4: 311-331.

Kamps, L. F., 1963. Mud distribution and land reclamation in the eastern Wadden shallows (mit Zusammenfassung). ILRI Publication 9, Wageningen.

Kauritsjeva, I. S. & I. D. Gromyko, 1974. Atlas pochv SSSR. (Russ., no summary). Kolos, Moscow.

Knibbe, M., 1969. Gleygronden in het dekzandgebied van Salland (with summary: Coversand gley soils in Salland, the Netherlands). Thesis Wageningen.

Kubiëna, W. L., 1953. Bestimmungsbuch und Systematik der Böden Europas. Enke Verlag, Stuttgart.

Kuźnicki, F., K. Konecka-Betley, A. Kowalkowski & S. Bialousz, 1974. Systematyka Gleb Polski (with summary: The classification system of Polish soils). Rocniki Gleboznawcze 25, 1. Pánstwowe Wydawnictwo Naukowe, Warszawa.

Lambert, Audrey, M., 1971. The making of the Dutch landscape, an historical geography of the Netherlands. Seminar Press, London/New York.

Maas, H. & E. Mückenhausen, 1970. Böden von Nordrhein-Westfalen, Maszstab 1:500 000 (Karte und Erläuterung zur Karte). Deutscher Planungsatlas, Bd. Nordrhein-Westfalen. Ak. Raumforschung und Landesplanung, Hannover, und Landesplanungsbehörde, Düsseldorf.

Mac Vicar, C. N., et al., 1977. Soil classification, a binomial system for South Africa. Dep. Agric. techn. Serv. Rep. S.A., Pretoria.

Miklazewski, S., 1924. Contribution à la connaissance des sols nommés 'rendzina's'. Comptes Rendus de la Conférence extraordinaire (IIIme Internationale) Agropédologique à Prague 1922: 312-318. Prague.

Mückenhausen, E., 1973. Pseudogleye und Gleye in der Bodengesellschaft der humiden, gemässigt warmen Klimaregion. In: Pseudogley and gley. Trans. Comm. V and VI of the ISSS: 147-157. Verlag Chemie GmbH, Weinheim/Bergstr.

Mückenhausen, E., 1975. Die Bodenkunde und ihre geologischen, geomorphologischen, mineralogischen und petrologischen Grundlagen. DLG-Verlag, Frankfurt am Main.

Mückenhausen, E. in Zusammenarbeit mit F. Heinrich, W. Laatsch & F. Vogel, 1962. Entstehung, Eigenschaften und Systematik der Böden der Bundesrepublik Deutschland. DLG-Verlag, Frankfurt am Main.

Mückenhausen, E. in Zusammenarbeit mit H.-P. Blume, F. Heinrich & S. Müller, 1977. Entstehung, Eigenschaften und Systematik der Böden der Bundesrepublik Deutschland. 2. Auflage. DLG-Verlag, Frankfurt am Main.

Müller, S., 1967. Südwestdeutsche Waldböden im Farbbild, nach Aufnahmen von K. Glatzel, R. Jahn u.a., unter Mitarbeit von G. Schlenker und J. Werner. Schriftenreihe der Landesforstverwaltung Baden-Württemberg. Bd. 23. Selbstverlag, Stuttgart.

Müller, W., 1954. Untersuchungen über die Bildung und die Eigenschaften von Knickschichten in Marschböden. Thesis Giessen.

Němeček, J. (Ed.), 1967. Průzkum zemědělských půd ČSSR, Díl 1 (summaries in Russ., Ger., and Eng.: Handbook of the large-scale mapping of agricultural soils in Czechoslovakia). Ministerstvo Zemědělství a Výživy, Praha.

Northcote, Keith H., 1971. A factual key for the recognition of Australian soils. 3d. ed. Rellim Techn. Publ., Glenside.

Orwin, C. S. & C. S. Orwin, 1967. The open fields. Clarendon Press, Oxford.

Paas, W., 1961. Rezente und fossile Böden auf niederrheinischen Terrassen und deren Deckschichten. In: Eiszeitalter und Gegenwart, Bd. 12: 165-320.

Pape, J. C., 1970. Plaggensoils in the Netherlands. Geoderma 4: 229-255.

Plaisance G. & A. Cailleux, 1958. Dictionnaire des sols. Maison Rustique, Paris.

Polunin, O., 1969. Flowers of Europe. Oxford Univ. Press, London.

Pons, L. J. & I. S. Zonneveld, 1965. Soil ripening and soil classification, initial soil formation of alluvial deposits with a classification of the resulting soils. Publication 13, ILRI, Wageningen.

Pons, L. J. & W. H. van der Molen, 1973. Soil genesis under dewatering regimes during 1000 years of polder development. Soil Sci. 116: 228-235.

Ragg, J. M. & B. Clayden, 1973. The classification of some British soils according to the comprehensive system of the United States, with contributions on micromorphology by P. Bullock. Soil Survey, Technical Monograph No. 3, Harpenden.

Rozov, N. N. & E. N. Ivanova, 1967. Classification of the soils of the USSR (Principles and a systematic list of soil groups). Soviet Soil Sci. 2: 147-155.

Schelling, J., 1955. Stuifzanden (with summary: Inland-dune soils). Detailed Reports of the Forest Research Station TNO, Bd. 2. Stichting voor Bodemkartering, Wageningen.

Schelling, J. & B. A. Marsman, 1973. Soil pattern and soil genesis in hydromorphic soils in brookvalleys. In: Pseudogley and gley. Trans. Comm. V and VI of the ISSS: 159-169. Verlag Chemie GmbH, Weinheim/Bergstr.

Seale, R. S., 1975. Soils of the Ely district (sheet 173). Memoirs of the Soil Survey of Great Britain. England and Wales, Harpenden.

Servant, J., 1973. Les sols des wateringues du Nord et du Pas de Calais. Inst. natn. Rech. agron., Service d'étude des sols, Montpellier.

Slicher van Bath, B., 1963. The agrarian history of western Europe, A.D. 500-1850. (Translated by Olive Ordishe). London.

Snacken, F., 1971. Les champs bombés du Pays de Waes. In: L'habitat et les paysages ruraux d'Europe. Comptes rendus du Symposium tenu à l'Université de Liège du 29 juin au 5 juillet 1969. Luik.

SSS (Soil Survey Staff), 1951. Soil survey manual. U.S. Dep. Agric. Handb. 18. U.S. Govt. Printing Office, Washington, D.C.

SSS, 1960. Soil classification, a comprehensive system, 7th approximation. Soil Conserv. Serv., U.S. Dep. Agric., U.S. Govt. Printing Office, Washington, D.C.

SSS, 1972. Soil series of the United States, Puerto Rico and the Virgin Islands: Their taxonomic classification. Soil Conserv. Serv., U.S. Dep. Agric., U.S. Govt. Printing Office, Washington, D.C.

SSS, 1975. Soil taxonomy, a basic system of soil classification for making and interpreting soil surveys. Soil Conserv. Serv., U.S. Dep. Agric. Handb. 436. U.S. Govt. Printing Office, Washington, D.C.

SSSA (Soil Science Society of America), 1968. The Marbut memorial slides, prepared and published by the Soil Science Society of America.

Staring, W. C. H., 1856. De bodem van Nederland, de zamenstelling en het ontstaan der gronden in Nederland. (Dutch). Kruseman, Haarlem.

Stefanovits, P., 1971. Brown forest soils of Hungary. Akadémiai Kiadó, Budapest.

Steur, G. G. L. & H. de Bakker, 1969. De bodemgesteldheid van het Veluwe-Randgebied (with summary: Soil conditions in the border area of the Veluwe). Basisrapport II in Het Veluwemeer. Rapp. en Meded. Zuiderzeewerken No. 7. Staatsdrukkerij, 's-Gravenhage.

Stichting voor Bodemkartering, 1961. Bodemkaart van Nederland, schaal 1:200 000, 11 bladen (Soil Map of the Netherlands, scale 1:200 000, 11 sheets, the English legend and the glossary are on sheet 11). Pudoc, Wageningen.

Stichting voor Bodemkartering, 1964 and following years. Bodemkaart van Nederland, schaal 1:50 000, in 110 bladen (Soil Map of the Netherlands, scale 1:50 000, in 110 sheets, in 1978 45 sheets were published; no English summaries). Pudoc, Wageningen.

Targulian, V. O. (Ed.), 1974. Arrangement, composition and genesis of sod-pale-podzolic soil derived from mantle loams; morphological investigation. Proc. Xth Int. Congr. Soil Sci., Moscow.

Tavernier, R. & R. Maréchal, 1958. Carte des associations de sols de la Belgique. Pédologie, Ghent 8: 134-182.

Ten Houte de Lange, S. M. (Ed.), 1977. Rapport van het Veluwe-onderzoek. Een onderzoek van natuur, landschap en cultuurhistorie ten behoeve van de ruimtelijke ordening en het recreatiebeleid (in Dutch). Pudoc, Wageningen.

Ter Wee, M. W., 1962. The Saalian glaciation in the Netherlands. In: Meded. Geol. St, Nwe Serie No 15: 57-76. Maastricht.

Thorp, J. & G. D. Smith, 1949. Higher categories of soil classification: Order, suborder, and great soil groups. Soil Sci. 67: 117-126.

USDA (United States Department of Agriculture), 1938. Soils and Men. Yearbook of Agriculture 1938. U.S. Govt. Printing Office, Washington, D.C.

Van den Broek, J. J. M. & L. van der Waals, 1967. The Late-Tertiary peneplain of South Limburg (The Netherlands). Silicification and fossil soils; a geological and pedological investigation. Soil Survey Papers 3, Soil Survey Institute, Wageningen.

Van den Broek, J. J. M. & H. W. van der Marel, 1968. Weathering, clay migration and podzolization in a hydromorphic loess soil. Geoderma 2: 121-150.

Van der Molen, W. H., 1957. The exchangeable cations in soils flooded with sea water. Thesis Wageningen.

Van der Voort, W. J. M., 1972. Ongestoorde gloeimonsters (with summary: Ignition of undisturbed soil samples, with one colour plate). Boor en Spade 18: 50-53. Veenman, Wageningen.

Van der Voort, W. J. M. & Chr. J. M. Kraanen, 1971. Toepassing van bronbemaling bij het maken van lakfilms (with summary: A well-point dewatering system for taking lacquer peels). Boor en Spade 17: 35-38. Veenman, Wageningen.

Van der Sluijs, P., 1970. Decalcification of marine clay soils connected with decalcification during silting. Geoderma 4: 209-227.

Van Giffen, A. E., 1964. De ouderdom onzer dijken (with summary: The age of our dikes). Kon. Ned. aardr. Gen., Tweede Reeks 81: 273-286.

Van Heesen, H. C., 1970. Presentation of the seasonal fluctuation of the water table on soil maps. Geoderma 4: 257-278.

Van Heuveln, B. & H. de Bakker, 1972. Soil-forming processes in Dutch peat soils with special reference to humus-illuviation. Proc. 4th Int. Peat Congr. I-IV: 289-297. Helsinki.

Van Veen, J., 1948. Grafieken van indijkingen in Nederland (Dutch). Kon. Ned. aardr. Gen., Tweede Reeks 65: 19-25.

Verhoeven, B., 1953. Over de zout- en vochthuishouding van geïnundeerde gronden (with summaries in Eng. and Fr.: Salt- and moisture conditions in soils flooded with seawater). Thesis Wageningen.

Visser, W. C., 1958. De landbouwwaterhuishouding van Nederland (with summaries in Eng, Fr., and Ger.: Agro-hydrological conditions in the Netherlands). COLN rapport No. 1.

Wiggers, A. J., 1955. De wording van het Noordoostpoldergebied, een onderzoek naar de physisch-geografische ontwikkeling van een sedimentair gebied (with summary: The genesis of the North-eastern polder area, a study on the physiographical development of a sedimentary region). Thesis Amsterdam.

Wurman, E., E. P. Whiteside & M. M. Mortland, 1959. Properties and genesis of finer textured subsoil bands in some sandy Michigan soils. Proc. Soil Sci. Soc. Am. 23: 135-143.

Zonneveld, I. S., 1960. De Brabantse Biesbosch, a study of soil and vegetation of a freshwater tidal area. Volume A, summary with text figures and tables. Thesis Wageningen.